Source Reduction and Waste Minimization

Source Reduction and Waste Minimization

Volume 2 of Advanced Zero Waste Tools: Present and Emerging Waste Management Practices

CHAUDHERY MUSTANSAR HUSSAIN

Adjunct Professor, Academic Advisor and Director of Chemistry and EVSc Labs, Department of Chemistry and Environmental Sciences, New Jersey Institute of Technology (NJIT), Newark, NJ, United States

MOSAE SELVAKUMAR PAULRAJ

Assistant Professor in Chemistry, Science and Math Program at the Asian University for Women (AUW), Chittagong, Bangladesh

SAMIHA NUZHAT

Junior Research Officer, Water and Life Bangladesh, Dhaka, Bangladesh

ELSEVIER

Elsevier
Radarweg 29, PO Box 211, 1000 AE Amsterdam, Netherlands
The Boulevard, Langford Lane, Kidlington, Oxford OX5 1GB, United Kingdom
50 Hampshire Street, 5th Floor, Cambridge, MA 02139, United States

Library of Congress Cataloging-in-Publication Data
A catalog record for this book is available from the Library of Congress

British Library Cataloguing-in-Publication Data
A catalogue record for this book is available from the British Library

ISBN: 978-0-12-824320-6

For information on all Elsevier publications
visit our website at https://www.elsevier.com/books-and-journals

Publisher: Candice Janco
Acquisitions Editor: Marisa LaFleur
Editorial Project Manager: Leticia M. Lima
Production Project Manager: Kumar Anbazhagan
Designer: Miles Hitchen

Typeset by Thomson Digital

Contents

About the Authors *xi*

1. Source reduction and waste minimization—concept, context, and its benefits **1**

1.1. Introduction 1
1.2. Source reduction and waste minimization 3
1.3. Commonly produced hazardous waste 4
1.4. Principles of sustainable waste minimization 5
1.4.1. 3R (reduce, reuse, and recycle) 5
1.4.2. Solid waste management 6
1.4.3. Liquid waste management 8
1.4.4. Integrated waste management strategies 10
1.4.5. Zero waste strategy 11
1.5. Co-benefits of waste source reduction and waste minimization 12
1.5.1. Environmental co-benefits 12
1.5.2. Health-related co-benefits 13
1.5.3. Economic benefits 15
1.5.4. Waste minimization and sustainable development goals (SDGs) 16
References 18

2. Source reduction, waste minimization, and cleaner technologies **23**

2.1. Introduction 23
2.2. Recycling 24
2.3. Composting 25
2.4. Biological treatment 26
2.4.1. Anaerobic digestion 27
2.4.2. Enzyme-based waste minimization 28
2.4.3. Membrane-based system 30
2.4.4. Waste autoclave 30
2.5. Chemical treatment 31
2.5.1. Electrolysis 32
2.5.2. Pyrolysis 33
2.6. Thermal treatment 34
2.6.1. Incineration 34
2.6.2. Molten salt oxidation 36
2.6.3. Hydrothermal carbonization 37
2.6.4. Hydrothermal liquefaction (HTL) 38
2.6.5. Plasma gasification system 39

2.7. Filtration 39
2.7.1. Biofiltration 40
2.7.2. Carbon filtering 40
2.7.3. Ultrafiltration 41
2.7.4. Reverse osmosis 42
2.7.5. Ocean cleanup system 43
2.8. Energy source modification 44
2.8.1. Renewable energy generation 44
2.8.2. Introducing alternative energy sources 45
2.9. Information technology-based waste minimization 46
2.9.1. RFID tracking 46
2.9.2. Apps and software 47
2.10. Automatic waste collection 47
2.11. Green chemistry: as a waste minimization strategy 48
References 52

3. Source reduction and waste minimization in electrical and electronics industry 61
3.1. Introduction 61
3.2. Commonly produced waste in the electrical and electronics industries: source and biogeochemical hazards 63
3.2.1. Metallic waste 64
3.2.2. Organic waste 67
3.3. Source-level waste minimization initiatives by the electrical and electronics industries 69
3.3.1. Process modification 70
3.3.2. Recycling 71
3.3.3. Laws, regulations, and policies intervention 78
References 80

4. Source reduction and waste minimization in healthcare industry 83
4.1. Introduction 83
4.2. Commonly produced waste in the healthcare industries: source and biogeochemical hazards 85
4.3. Source-level waste minimization initiatives by the healthcare industries 85
4.3.1. Waste segregation 91
4.3.2. Product substitution, process modification, and technological advancement 93
4.3.3. Reducing, reusing, and recycling 98
4.3.4. Thermal treatment 101

4.3.5. Biochemical treatment 102
4.3.6. Laws, regulations, and policies intervention 103
References 107

5. Source reduction and waste minimization in construction industry 111
5.1. Introduction 111
5.2. Commonly produced waste in the construction industries: source and biogeochemical hazards 113
5.3. Source-level waste minimization initiatives by the construction countries 114
5.3.1. Product substitution, process modifications, and technological advancement 119
5.3.2. 3R approach 121
5.3.3. Laws, regulations, and policies interventions 123
References 125

6. Source reduction and waste minimization in chemical industry 127
6.1. Introduction 127
6.2. Commonly produced waste in the chemical industries: source and biogeochemical hazards 129
6.2.1. Solid chemical waste 129
6.2.2. Liquid chemical waste 129
6.2.3. Gaseous chemical waste 130
6.3. Source-level waste minimization initiatives by the chemical industries 131
6.3.1. Product substitution, process modification, and technological advancement 131
6.3.2. 3R approach 132
6.3.3. Prerelease treatment 133
6.3.4. Laws, regulations, and policies interventions 133
References 134

7. Source reduction and waste minimization in food industry 137
7.1. Introduction 137
7.2. Commonly produced waste in the food industries: source and biogeochemical hazards 139
7.3. Source-level waste minimization initiatives by the food industries 139
7.3.1. Process modifications 143
7.3.2. 3R approach 144
7.3.3. Laws, regulations, and policies interventions 146
References 147

8. Source reduction and waste minimization in energy production industry **149**
8.1. Introduction 149
8.2. Commonly produced waste in the energy production industries: source and biogeochemical hazards 151
8.3. Source-level waste minimization initiatives by the energy producing industries 151
8.3.1. Product substitution 151
8.3.2. Technological efficiency enhancement 154
8.3.3. Recycling and reusing 155
8.3.4. Laws, regulations, and policies interventions 156
References 156

9. Source reduction and waste minimization in textile industries **159**
9.1. Introduction 159
9.2. Commonly produced waste in the textile industries: source and biogeochemical hazards 160
9.3. Source-level waste minimization initiatives by the textile industries 163
9.3.1. Product substitution 163
9.3.2. Technological efficiency enhancement 164
9.3.3. Reduce, reuse, and recycling 165
9.3.4. Laws, regulations, and policies interventions 167
References 167

10. Source reduction and waste minimization in the mining industries **169**
10.1. Introduction 169
10.2. Commonly produced waste in the mining industries: source and biogeochemical hazards 170
10.3. Source-level waste minimization initiatives by the mining industries 173
10.3.1. Product substitution 173
10.3.2. Technological efficiency enhancement 174
10.3.3. Inclusion of 3R approach 175
10.3.4. Laws, regulations, and policy interventions 175
References 176

11. Challenges and recommendations for source reduction and waste minimization **177**
11.1. Introduction 177
11.2. Corresponding challenges and recommendations for source reduction and waste minimization 177

11.2.1. Industrial reluctance 178
11.2.2. Inadequate training facilities 179
11.2.3. Technological limitations 180
11.2.4. Financial limitations 181
References 181

Index 183

About the Authors

Dr. Chaudhery Mustansar Hussain

Chaudhery Mustansar Hussain, PhD, is an Adjunct Professor, Academic Advisor, and Lab Director in the Department of Chemistry and Environmental Sciences at the New Jersey Institute ofTechnology (NJIT), Newark, United States. His research is focused on environmental management, nanotechnology, advanced materials, and analytical chemistry of various industries. Dr. Hussain is the author of numerous papers in peer-reviewed journals as well as a prolific author and editor of scientific monographs and handbooks in his research areas for Elsevier, Royal Society of Chemistry, CRC, Springer, and others.

Dr. Mosae Selvakumar Paulraj

Dr. Mosae Selvakumar Paulraj received his Bachelors and Masters degree in Chemistry from Manonmaniam Sundaranar University, Tamil Nadu, India. He is a diploma holder in teacher education. He started his professional career in Jubilant Biosys Ltd., Bangalore, as a scientist in the cheminformatics division. Then he joined as a research fellow at Central Salt And Marine Chemical Research Institute (CSMCRI)—a unit of the Council of Scientific and Industrial Research, and obtained his PhD degree. He also worked as a DAAD research fellow in Technical University, Kaiserslautern, Germany. Presently, Dr. Mosae is working as Assistant Professor of chemistry at Asian University for Women (AUW), Chattogram, Bangladesh. Before joining AUW, Dr. Mosae has worked as Assistant Professor of chemistry at Karunya University, Coimbatore, Tamil Nadu, India. Dr. Mosae's research interests include synthesis of small bioactive molecules, development of value-added products from Asian palm (Palmyraculture), molecular recognition/machines, development of sensor molecules for anions/cations/small molecules in water, water treatment, chemo/biosensors, host-guest chemistry, DNA binding-drug delivery using nanoparticles-material chemistry, antibacterial and wound healing molecules, photochemistry, metallo-supramolecular material chemistry-helicate, MOF, flame retardant materials, green chemistry, waste to wealth, pollution control, environmental conservation, science for self-reliant lifestyle/sustainable development, grassroots innovations, and social entrepreneurship.

Samiha Nuzhat
Samiha Nuzhat completed her undergraduate studies with a double major in Environmental Science and Bioinformatics from Asian University for Women (AUW), Chattogram, Bangladesh. Currently, she is working as a Junior Research Officer at Water and Life Bangladesh, a French NGO dedicated to assure safe and reliable domestic water service, sustainable sanitation, hygiene training, and other need-based community empowerment initiatives, specified for the urban slum dwellers. Samiha also worked a Teaching Assistant (TA) for two and half years and as a TA, she assisted other undergraduate students in understanding the concepts of multiple science courses. She is also a passionate researcher who has been working in multiple research projects related to the fields of environment and sustainability. Samiha has diversified research experience involving different vulnerable communities (refugees, homeless people, slum dwellers, etc.) on different issues. Her major research interests encompass water and waste management, green technology, WASH (Water, Sanitation, and Hygiene), GIS (Geographic Information System), remote sensing, and other emerging fields of environmental research.

CHAPTER 1

Source reduction and waste minimization—concept, context, and its benefits

1.1 Introduction

As wise people say "Prevention is better than cure," it is substantially applicable in case of waste management strategies that recommends to prevent its generation instead of its treatment after being generated. This is where the concept of "Waste Minimization" comes which suggests suitable approaches to reduce amount of waste produced (Clark, 2012). It is noted that despite the implementation of many waste management initiatives, production of waste materials are increasing day by day due to unprecedented urban expansion and industrial growth (Song et al., 2015). Therefore, it is being harder to treat produced waste with limited waste management facilities. In this regard, waste minimization strategies at source-level can ensure gradual reduction in waste production and subsidize successful implementation of other waste management strategies (Cheremisinoff, 2003). This will also assure minimized exposure led minimum hazards since waste is reduced before its environmental exposure. Thus waste minimization approaches ensure efficient utilization of limited treatment facilities and limited resources while managing the produced waste. Unlike many traditionally followed waste management strategies, waste minimization, and source reduction approaches focus on the root causes of excessive waste generation. As such approaches are adopted and implemented at the source-level of waste generation, these are found to be less harmful and comparatively inexpensive than many complex waste management strategies (Cheremisinoff, 2003). This makes waste minimization approaches more preferable over many other waste management strategies.

Unfortunately, along with the undeniable benefits that waste minimization strategies can bring about, it has been a matter of huge controversy due to its complex nature. Successful implementation of waste minimization strategies requires constant reviewing, monitoring, evaluation, and reporting (Zorpas and Lasaridi, 2013). Due to such complexities, many

Source Reduction and Waste Minimization
http://dx.doi.org/10.1016/B978-0-12-824320-6.00001-0

countries are struggling to design suitable strategies that will minimize waste-related issues drastically. As a result, frequency and occurrence of hazardous waste-borne diseases, environmental degradation, and occurrence of other harmful incidences are increasing through increased exposure to waste materials along with production of increased waste. This problem is even more complex due to the huge diversity of the waste issue (Chaaban, 2001). It may cause many troubles that might not be realized instantly. To minimize extent of this issue, active participation of non-government agencies along with government agencies, is required. Especially the major producers of waste materials should take a leading role in minimizing the share of produced waste.

As the industries are the major responsible agencies that produce huge share of waste materials every year, they should be responsible to contribute actively in managing the waste materials. Nowadays, many of the industries are trying to adopt suitable measurements to treat waste produced at various steps of the industries. Unfortunately, in many cases, the industries fail to treat their waste sustainably; mostly because of reluctance of industrial authority, increased cost of waste treatment, technological unavailability, huge diversity of produced waste, and for so many other reasons (Heidrich et al., 2009; Chen et al., 2010). Even if policies are being adopted to ensure chemical and biological treatment of industrial waste before disposal, there are evidences that many of the industrial wastes are disposed without proper treatment (Kazi et al., 2005). In this regard, the condition of the developing countries are worse than the developed countries due to their limitations of technologies, policy adoption, and financing agencies that strictly prohibit untreated waste disposal by the industries (Dhokhikah and Trihadiningrum, 2012). If source-level waste minimization strategies are introduced, these industries are hoped to ensure successful implementation of the waste minimization strategies due to its comparatively easier means of implementation than other waste management strategies.

This chapter defines source reduction and waste minimization and then classifies the commonly produced hazardous waste materials which should be minimized. It also explores some widely used waste minimization strategies such as 3R approach, solid and liquid waste management, and integrated waste management strategies to elaborate the current extent of initiatives adopted to minimize this issue. Then it discusses the major co-benefits of waste minimization strategies that can be brought about in multiple sectors.

1.2 Source reduction and waste minimization

Waste minimization is the concept that discusses on approaches to reduce the amount of hazardous waste generation. Waste minimization is regarded as a waste management strategy, which specifically focus on minimization of waste production instead of encouraging postproduction waste treatment (Cheremisinoff, 2003). Waste minimization encourages less waste production by modifying processes involved with waste source. On that account, source reduction denotes source-level mitigation of waste materials; in other words, source reduction suggests waste treatment before its exposure to the environment. At an industrial level, waste source reduction and waste minimization can be achieved through process and product modification, constant monitoring and recycling (Babu et al., 2009). On the contrary, at a household level of waste source reduction, public participation can play the key role (Lober, 1996). As the industries are the major producers of hazardous waste, source reduction at an industrial setting is more urgent as well as more complex than household-level waste source reduction. Considering the diversity of waste materials produced in the industries, process modification of the industries has been an urgent need to reduce industrial waste source of hazardous waste (Chaaban, 2001). Through process modification, the industries often recover many useful resources which can subsidize the cost of source reduction implementation. Apart from this, the greater benefits of waste minimization and source reduction will include environmental protection, minimization of waste-borne health hazards, and so on.

There is a lack of data and limited initiatives for large scale waste minimization practice. Especially the low income countries hardly prioritize waste minimization over economic growth (Van Berkel, 2004). Besides, though the industrial revolution suddenly boosted up waste generation, such waste minimization initiatives didn't increase simultaneously. There is yet a scarcity of industrial initiatives to minimize waste generation through process modifications. Fortunately, many developed countries are adopting policies to minimize waste production that aims sincere initiatives of the waste producers to minimize the amount of produced waste. This is why it is hoped that one of the major concerns of next industrial revolution would be environmental sustainability (Graham, 2000). However, despite multiple initiatives, waste minimization rate could not yet reach to a safer limit. World Bank (2019) predicts an increase of 70% waste generation by 2050. It denotes how insufficient co-current waste minimization initiatives are. Besides, nearly 90% of the solid wastes are dumped openly without any sort of treatment that contributes to spread deadly diseases, climate change, land infertility, poverty, urban violence, and many more unexpected events (World Bank, 2019). Altogether, not a very optimistic scenario is noticed in terms of the future of large scale waste minimization and source reduction.

1.3 Commonly produced hazardous waste

Hazardous waste materials are those waste materials that carry the characteristics of physical, chemical, and biological hazards and these materials are commonly known as harmful for our health and environment. Here, hazard denotes presence of harmful agents that can cause danger or risk in later terms. Likewise, hazardous waste materials can contain harmful agents of different types. Environmental Protection Agency (EPA) classifies hazardous waste into three major classes; listed waste, characteristic waste, and mixed radiological waste (EPA, 2020). Listed wastes are those that have been considered under the list of hazardous waste by Code of Federal Regulations (CFR) in section 261. Characteristic wastes have characteristics of ignitability, corrosively, reactivity, and toxicity. And mixed radiological wastes are those wastes that contain radioactive materials. Shortly speaking, whatever classification method is followed, hazardous waste materials are those which have the potential to bring about negative impacts on human health and environment. Here, types and variety of such hazardous wastes vary largely from one industry to another industry. For instance, while metallic waste is the major class of electrical and electronics industry generated waste, in case of healthcare industry infectious waste is the major category of generated waste.

Realizing the huge diversity of hazardous waste materials, it is understood that from household waste to clinical waste, almost every waste material can be hazardous waste based on the components it contain. In the households, dust, organic waste, sanitary waste, etc. are daily produced hazardous waste that can lead to harmful diseases when people are exposed to these. Similarly, industrial waste materials, harmful chemicals such as dyes, acids etc., by-products like carbon dioxide, sulfur dioxide, etc., and microorganisms, can be regarded as hazardous waste materials. Along with these, untreated industrial waste disposal can also form hazardous waste materials later when released in environment (Malviya and Chaudhary, 2006). Apart from disposal, during some other steps such as product transportation, product processing, etc., such hazardous chemicals may get exposed to human and environment. This is why realizing the detrimental impacts that such waste materials may cause, regulations are being introduced so that the industries and respective agencies take a leading role in minimizing exposure to hazardous waste.

1.4 Principles of sustainable waste minimization

Sustainable waste minimization strategies work under several principles. Based on the purpose of a waste management program, a suitable principle or strategy is chosen. Largely, waste management strategies can be classified in two major classes; solid waste management and liquid waste management. There are significant share of overlapping strategies that are commonly found in both of these classes. Some of such strategies are 3R approach, integrated waste management strategies, and so on. And waste minimization strategies can also be effective while implementing principles of these popular waste management strategies. However, before implementing such strategies there are several criteria to be considered. This includes geographic suitability, environmental impact analysis, cost-benefit analysis, cultural suitability, and many more (Chi et al., 2020; Javaheri et al., 2006). Otherwise, there are chances of strategy failure in short- or long-term. This chapter briefly discusses some major features of several sustainable waste minimization strategies commonly implemented in different contexts.

1.4.1 3R (reduce, reuse, and recycle)

3R is a popular concept of waste management that encourages reuse, recycling, and reduction of waste. 3R is an important component of integrated solid and liquid waste management mechanism (Memon, 2010). The major purpose of 3R in waste management is to produce useful resources out of waste materials. Based of the available resources and technologies, 3R concept can be implemented both in large scale and in small scale bringing proper modifications and contextualization. Implementation of different 3R approaches helps to recover various types of useful resources from waste materials. At the same time, 3R ensures the best use of resources before throwing them away.

To understand how 3R approach works in brief, Fig. 1.1 can illustrate a chain of the workflow followed under this approach. Though all these steps are inter-connected, usually the basic workflow functions in the below-mentioned direction.

Applications of 3R approach are quite vast. Valuable metal isolation from waste materials using 3R approach is found effective in certain settings

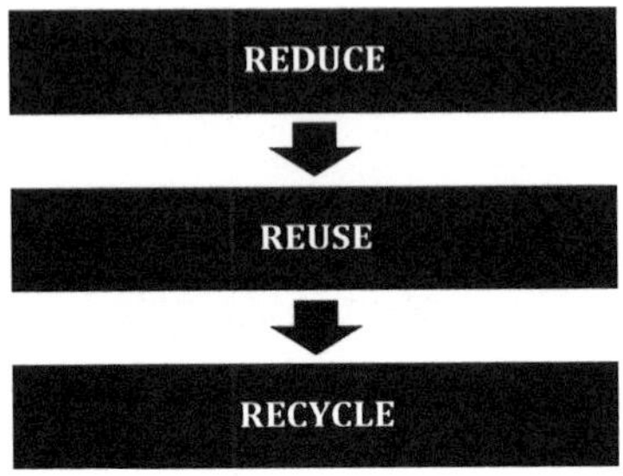

Figure 1.1 ***Basic workflow of 3R model.***

which lead to significant share of iron, silicon, and other metal recovery (Aadal et al., 2013). Most of the organic fertilizers are produced from organic waste materials (Obi et al., 2016). Especially for agricultural waste management, 3R helps to reduce agricultural waste as well as produce useful fertilizers that are found to be a successful in agriculture based localities (Agamuthu, 2009). 3R is also appropriate for greenery restoration in natural setting (Tangwanichagapong et al., 2017). Additionally, it was found that using 3R approach, construction industries of China could demolish their production costs (Huang et al., 2018). Besides, inclusion of 3R studies as a part of school curriculum helped the school going kids to enhance their environmental knowledge and it also helped them to reduce, reuse, and recycle the solid waste produced in the schools (Mahat et al., 2016). Road construction using solid wastes is another significant example of 3R applications. Despite some environmental drawbacks of this, it is found effective in places where there is lack of technologies to adopt highly efficient 3R approaches (Khursheed and Singh, 2017). Thus, 3R approach is regarded as a sustainable medium of waste management in different contexts.

1.4.2 Solid waste management

Solid waste management is one of the most important areas of waste management that requires huge effort to be successfully implemented. Solid wastes are those waste materials that have physical properties of solid substances. Household waste, old papers, dysfunctional instruments, etc., are examples of commonly produced solid wastes. Due to the huge amount of produced solid waste everyday around the corners of the world, it is an alarming issue to be tackled by the concerned authority. Usually after collecting the solid wastes, those waste materials are segregated primarily and then based on the types and available facilities, some of the waste materials are processed to reuse or recycle; some are burnt if incineration facility is available and some are disposed in landfills after basic treatment or without any sort of treatments (Arena et al., 2003; McDougall et al., 2008). A proper solid waste management system is required for environmental welfare, improved lifestyle, economic viability, and for overall sustainability of a society (Gentil et al., 2009; Ramachandra and Varghese, 2003).

Solid waste management incorporates steps of different tasks under its major workflow which have been mentioned in Fig. 1.2 briefly. But based on the availability of resources and technology as well as, according to the interest of the respective project authority, the workflow might differ from context to context through addition of elimination of some steps.

Solid waste management is one of the most crucial component of every city governance worldwide because of the dreadful consequences that might be caused due to the municipal government's inaction on this matter (Hoornweg and Bhada-Tata, 2012). Especially in case of the cities from developing countries, the city authorities often fail to introduce effective initiatives in terms of municipal solid waste management due to the increasing amount of produced waste, lack of experienced manpower, insufficient land resources, and high cost associated with the issue (Guerrero et al., 2013). In addition to it, failure to cope up with technology advancement for solid waste management is another arena that significantly lags behind the cities of developing countries (Christensen, 2011). However, there is evidence that with fruitful policy adoption and government initiatives, countries like Malaysia have been quite successful to manage their city's solid waste under the regulation of Ministry of Housing and Local Government along with the involvement of assigned private sectors (AbdManaf et al., 2009). On the contrary many countries like China and India are suffering miserably to cope up with the increasing share of municipal solid waste produced

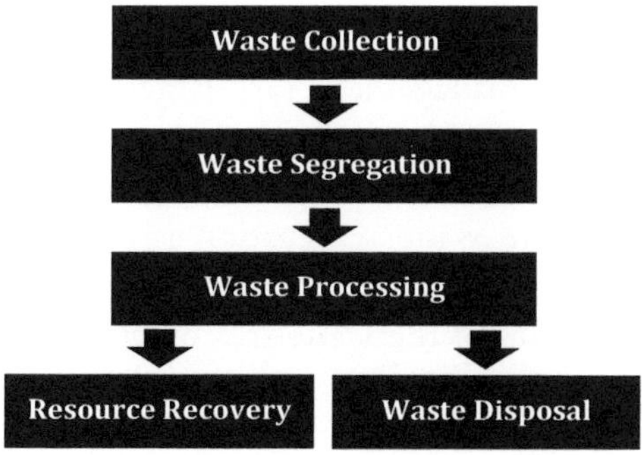

Figure 1.2 ***Basic workflow followed in solid waste management.***

every year (Sharholy et al., 2008; Zhang et al., 2010). Despite undertaking several initiatives, the countries often fail to manage their solid waste due to the insufficiency of facility and indifference of the citizens (Sharholy et al., 2008). Besides, dumping waste materials without proper treatment is another common practice for the municipal authority that requires sincere initiative of the rulers to be solved (Vasanthi et al., 2008). Additionally, regulatory frameworks introducing multi-faceted trade-offs, technological progress and economic policy instruments are needed to put an end to improper municipal solid waste management (Pires et al., 2011). Thus, multiple solid waste management strategies altogether work to ensure proper solid waste management in municipalities.

1.4.3 Liquid waste management

Liquid waste management is another major area of waste management that requires explicit interventions of the authority to be managed properly. Though harm caused by exposure to waste materials are determined by multiple factors, usually liquid wastes are more toxic than solid wastes due to its potential of easier intake by human body (Feng et al., 2005). Besides, because of its fluid nature, it is often tough to manage and treat liquid waste materials without the interventions of sophisticated tools and mechanisms (Dunn and Guerra, 2005). Among the various types of liquid waste, some of the most harmful ones are sanitary waste, dairy waste, liquid industrial waste, laboratory wastes, and so on. Reduction of liquid waste source requires installation of sophisticated system with multi-layered treatment facilities (Al Yaqout, 2003). Unfortunately many of the low-income developing countries lag behind in adopting suitable policy and technical settings to manage their liquid waste sustainably (Wilson et al., 2006). Realizing the huge diversity and increasing quantity of produced liquid waste materials, it is necessary to undertake suitable measurements to minimize waste sources of liquid waste.

Though liquid waste management vary largely from different types of waste produced, there are some commonly recommended basic steps to be implemented for several liquid waste management initiatives. Unlike solid waste management where waste segregation is an important step, liquid waste management requires filtration instead of waste segregation. Then based on available facilities of the waste treatment plant, different physical, chemical and biological treatments are conducted. Fig. 1.3 illustrates a basic workflow commonly implemented for several liquid waste minimization contexts.

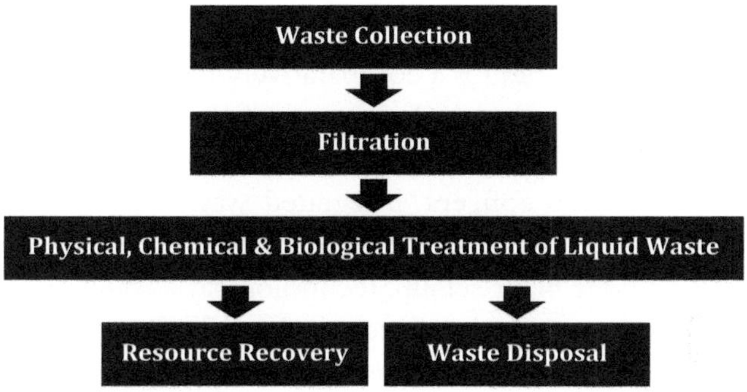

Figure 1.3 ***Basic workflow followed for liquid waste management.***

Liquid waste management, being an integral concern of both the government and the non-government agencies, requires both intervention of suitable policy implementation and technological advancement to reduce the dreadful impacts of improper liquid waste management strategies on individuals and on the environment. There are multi-layered risks involved in the whole cycle of liquid waste management; especially during collection, transportation, and disposal of the waste, the risks are higher (Rumph, 2001). The impacts are not identical for different kinds of liquid waste. For example, among many different types of liquid waste, radioactive liquid waste materials are currently regarded as a very harmful class of liquid waste. Unfortunately, there are lack of enough efficient treatment technologies to treat radioactive waste materials causing severe health conditions when exposed to such waste materials (Borai et al., 2009). Another significant type of liquid waste class, clinical liquid waste, are often disposed in the sewer system without any sort of treatment that leads to exposure to hazardous chemicals (Abdulla et al., 2008). Therefore, to reduce exposure to such harmful chemicals, it is necessary to design waste treatment plants that will treat liquid waste sustainably before disposal. In this regard, the purposes of different treatment plants are decided by the availability of waste treatment facilities. In many cases, environmental protection is the major purpose while treating liquid waste; for which reason, the waste materials are treated to a certain level so that some certain indicators are at a safe range for ecology and environment (Tóth, 2015). In some cases, resource recovery is the major concern for which some certain types of resources (metallic, organic, etc.) are isolated from the liquid waste (Titus and Surma, 2001). In many cases, such treatment plants are subsidized by the government to ensure proper management of the liquid waste.

1.4.4 Integrated waste management strategies

Integrated waste management is a comparatively newer concept in the field of waste management that optimizes many of the former waste management practices and integrates new dimensions of waste management strategies (Seadon, 2006). The concept, integrated waste management, involves both the technical and non-technical components of waste management; where waste treatment plant set-up, technological advancement, chemical treatment, etc. are components of technical attributes of integrated waste management and policy adoption, awareness campaigns, etc. are part of non-technical components of integrated waste management (Harremoës, 1997). Nowadays, integrated waste management is an emerging field of waste management, as adoption of certain technologies cannot solve the waste-related issues sustainably, unless other suitable non-technical measurements are not adopted simultaneously. However, as the extent of integrated waste management strategies is expanding day by day, the complexity involved with smooth implementation of integrated waste management strategies is also increasing gradually causing an urgent need to design a sustainable framework of effective integrated waste management strategies (Seadon, 2006). Successful implementation of integrated waste management strategies, therefore, requires integration of multiple stakeholders from different fields.

Integrated waste management strategies can be significantly beneficial and impactful in multiple ways comparing with the outcomes of traditional isolated initiatives of waste management. Some basic components of integrated waste management are illustrated in Fig. 1.4. Based on the extent of available facilities, integrated waste management set-up alters in different contexts. As per the authoritarian interest, both long- and short-term

Figure 1.4 ***Technical and non-technical components of integrated waste management.***

outcomes can be targeted under integrated waste management system. For instance, implementation of integrated waste management mechanisms can be regarded as a climate change stabilization wedge while suitable climate change mitigation technologies are adopted (Bahor et al., 2009) that will have long-term outcome of such integrated waste management strategies. On the other hand, if recycling or other short-term strategies are adopted as a part of such waste management strategy, the benefit out of this strategy will be achieved in a short-term (Blengini et al., 2012). To implement such strategies, it is necessary to analyze the life cycle of the waste materials. Through life cycle analysis of waste, management strategies can be designed more effectively. Besides, the effectiveness of such waste management strategies can be determined through waste life cycle analysis (Seadon, 2010). Among many of the positive outcomes of such waste management strategies, it was found that efficiency in recycling is multiplied by integrating various components of integrated waste management (Blengini et al., 2012). Even resource recovery rate increases by adopting integrated waste management strategies (Giugliano et al., 2011). Thus, integrated waste management strategies can have multiple benefits as well as, it can multiply the usual benefit of any waste management strategy.

1.4.5 Zero waste strategy

Zero waste concept denotes certain modifications based on life cycle analysis of all the resources so that all products are reused and recycled (Song et al., 2015). It ensures a significant share of resource savings under zero waste concept. The basic concept and objective of both zero waste and waste source reduction is to minimize waste generation as much as possible. The major difference is that zero waste concept focuses on no waste production at all while source reduction focuses on waste minimization at the source-level or before exposure to the environment. Adopting zero waste strategies can be beneficial to implement waste minimization and source reduction more readily. Though being a completely zero waste city is nearly impossible in reality cities like San Francisco, Adelaide, etc. have gained immense success in term of the city waste minimization adopting zero waste strategies (Song et al., 2015). Here, zero waste strategies require availability of recycling facilities and treatment facilities for almost all the commonly generated produced waste. Besides, the all the community members and the local authorities should be aware of the zero waste concept and its benefits so that all the community members can sincerely contribute in achieving zero waste locality (Cole et al., 2014). Efficient household waste collection and industrial waste disposal are two important areas in this regard. It is because, negligence in these sectors might ruin all the positive outcomes brought by zero waste strategies (Zaman, 2015). This is why the industries and the waste collection authority should formulate their strategies, plans, and programs to implement zero waste concept. Unless all the stakeholders' full effort is ensured to make a zero waste locality, the outcomes might not be as expected.

There are several challenges and bindings related to the zero waste concept which might hamper its implementation despite the sincere effort of the stakeholders. Technological unavailability and huge cost of project implementation might be a hindrance in implementing zero waste concept in reality (Song et al., 2015). Besides, socioeconomic situation of a society largely controls success of a new strategy adoption in a society. Therefore, before implementing any strategy, its acceptance potential should be analyzed. Based on such analysis, suitable zero waste strategies should be designed accordingly. Thus, considering both the potentials and corresponding challenges of zero waste strategy, certain zero waste initiatives implementation can help in achieving significant reduction in waste generation.

1.5 Co-benefits of waste source reduction and waste minimization

Waste source reduction and waste minimization strategies bring about multiple benefits for environment, health, and economic sector. Most of these benefits are involved with waste separation and resource recovery (Menikpura et al., 2013; Nguyen et al., 2015). In such processes, waste-borne risks are minimized and at the same time, valuable resources are often extracted. Some of the major ways through which different sectors can be benefitted from source reduction and waste minimization processes have been discussed later.

1.5.1 Environmental co-benefits

Waste minimization strategies are required to protect our environment, as untreated waste can be the core reason behind the rise of many environmental issues. Presence of increased toxic elements causing harm to both the biotic and abiotic elements of the environment is associated with increased urbanization-led waste disposal (Simon et al., 2016). Some of the major environmental impacts of such improper waste disposal mechanisms include damaged biodiversity, air and water pollution, etc. (López-González et al., 2015; Hung et al., 2012). At the same time, despite having significant waste reduction potential, waste management strategies such as incineration, landfills, etc. can also damage carbon balance of our environment (Nabavi-Pelesaraei et al., 2017). However, apart from the direct major impacts of unsustainable waste management system on the environment, it can cause many dreadful impacts through complex pathways. This is why, implementing sustainable source reduction and waste minimization strategies will bring greater benefits for the environment.

Implementation of suitable waste management strategies help to improve environmental parameters that contribute to the environmental co-benefits of source reduction and waste minimization strategies. First, large scale implementation of waste source reduction mechanisms will help to protect biodiversity by upgrading environmental parameters in a suitable range minimizing waste-borne pollution. Ecological balance of species distribution is maintained properly when environmental parameters are at a suitable range (Domene et al., 2008). Aquatic flora and fauna are preserved while waste-borne water pollution is minimized (Goel, 2006). Useful microbes can grow in presence of low waste, as chemical wastes are found to be harmful for the growth of different essential microorganisms (Ahmann et al., 1994). Besides, sustainable waste source reduction system can also be useful to decrease rate of air pollution, a significant source of global climate change. Based on the types of waste source, mechanism of air pollution might be different. For instance, radioactive wastes can release radioactive materials in our environment for which if the disposal of radioactive waste is controlled properly, waste-borne air pollution will be reduced as well (Texier et al., 2003). Other organic waste materials can also release carbon dioxide and methane while decomposing (Abdelmalek and Rutsch, 1997). Therefore, if more regulations to minimize such waste sources are ensured, air pollution and its contribution to global climate change will also be minimized.

1.5.2 Health-related co-benefits

Several health benefits are achieved when waste sources are reduced, as ineffective waste management strategies degrade health conditions in various ways. Waste management related health issues are caused due to adoption of no waste management strategies and due to the inefficiency of existing waste management strategies. Therefore, to gain maximum health-benefits out of waste source reduction initiatives, it is necessary to adopt both suitable waste management strategies and to modify existing waste management strategies.

Waste sources are significantly responsible for multiple health hazards for which sustainable waste source reduction mechanisms will help to minimize the occurrence and severity of such health issues. Different types of waste sources are responsible for different health issues. For instance, exposure to electronic waste may cause thyroid problem, stillbirth, premature growth, and many more developmental problems by releasing neuro-toxicants in environment (Grant et al., 2013). Therefore, if e-wastes are treated sustainably, it is possible to avoid many of the earlier mentioned e-waste caused health issues. On the other hand, another major waste class, clinical waste are highly infectious which can lead inflammation and infections in multiple organs while disposed without proper regulations (Hossain et al., 2011). Untreated wastes have the potential to cause hepatitis E infection and early aging; especially when such wastes are mixed with food source and water source (Ceylan et al., 2003). This is why if proper treatment of waste sources are ensured before disposal, many of the earlier-cited health issues will be minimized bringing significant co-benefits of waste minimization in health sector.

Apart from the direct health benefits gained from waste management strategies, it is necessary to modify traditional waste management processes to minimize health hazards. In this regard, incineration and landfills can be two major examples of unsustainable waste management mechanisms that require further modifications to minimize their negative impacts. There are evidences that people living nearby incineration and landfill-closed sites are more prevalent to reproductive and developmental issues, cancer incidence, and higher mortality (Rushton, 2003). Even workers working with less protection in waste management sites are often exposed to respiratory and musculoskeletal problems (Rushton, 2003). Epidemiological studies have proved that exposure of heavy metals and radioactive wastes can amplify occurrence of cancer and gastrointestinal problems (Giusti, 2009). This is why technological improvisations of such processes are required which will help to reduce many of the mentioned health hazards. Besides, it was found that exposure to solid waste frequently is responsible for respiratory infection, inflammation and acute diarrhea (Athanasiou et al., 2010; Reyes et al., 1997). If the workers involved in such processes are provided with protective outfits, they will be able to avoid such occupational hazards. Thus, both new sustainable waste management strategies and modifications of existing waste management strategies will help to ensure multiple health benefits.

1.5.3 Economic benefits

Though it is commonly considered that waste reduction mechanisms require huge investment for smooth operation and maintenance, when its corresponding economics related co-benefits are considered, it can be found that waste minimization initiatives can be economically beneficial. As it is mentioned earlier, lack of efficient waste management initiatives can lead to harmful consequences on people's health and on our environment, likewise, it can lead to a huge loss for a country's economy. It is necessary to critically analyze the economy of waste management, as it is related with cost reduction for health and environment related expenditure (Porter, 2010). It is found that households with a proper waste management systems need to spend less than households with improper waste management system for medicine purpose; at the same time, total expenditure for public health programs increases when waste management systems are not upgraded enough (Martuzzi et al., 2010). Besides, variance of socioeconomic status controls frequency of exposure to waste source that finally determines occurrence of various health issues (Martuzzi et al., 2010). Even environmental degradation caused by lack of waste management strategies lead to industrial process inefficiency or less production, health hazards, and so many other problems that require higher expenditure in later terms (Soukopová and Malý, 2013). Thus, waste management strategies are related with economics through different variables like health and environment.

Establishment of a proper waste management system can lead to economic benefit through production of essential goods or cost minimization. Waste minimization reduces harm caused in health sector and on our environment which reduces expenditure for such purposes. Likewise, through waste minimization processes, essential goods are often produced which have market value. Recycling is one of the major ways through which waste management strategy can produce valuable goods. Though recycling process can charge some money, the produced products are mostly more valuable than the process cost (Porter, 2010). Similarly, reducing and reusing can also be profitable. Among the economics related co-benefits of other waste management strategies, waste mediated energy production can be a noteworthy example. In this process not only the produced waste is treated but also one of our major need, energy is produced. Jamasb and Nepal (2010) argue that despite the challenges involved in this process, an explanatory cost-benefit analysis conducted in the context of United Kingdom found out the huge potential and forthcoming benefit of this process. Another study by Weng and Fujiwara (2011) established a financial model of municipal solid waste management system that is found to be profitable in terms of its cost-benefit analysis. Most importantly, though waste management strategies might not seem much profitable in short term, it is a sustainable means of economic benefit in longer terms (Aydemir-Karadag, 2018). Therefore, designing a sustainable waste management system is hoped to be economically beneficial.

1.5.4 Waste minimization and sustainable development goals (SDGs)

Waste minimization initiatives can directly or indirectly help a nation to achieve several targets of sustainable development goals (SDGs). There are several ways through which waste minimization strategies can facilitate achievement of SDG's targets. First, waste minimization initiatives are needed for environmental sustainability (NA et al., 2020). Unless environmental sustainability is achieved, city growth cannot be sustainable for which environmental welfare of urban areas has been regarded as an important target under SDG 11 or Sustainable Cities and Communities. In this regard, introducing proper waste minimization strategies in the city areas should be taken into consideration. Second, as the SDG 7 (Affordable and Clean Energy) encourages introducing cleaner energy, implementation of waste to energy concept which is a clean waste minimization practice can be regarded in achieving SDG 7. Potential of waste to energy concept is found quite appropriate in gaining target of SDG 7, though there are needs for more technological advancement (AlQattan et al., 2018). Third, as it has been discussed earlier that waste minimization can bring about multiple benefits in health and economics sector, suitable waste minimization initiatives will help to attain several targets under SDG 3 (Good Health and Wellbeing) and SDG 8 (Decent Work and Economic Growth). It denotes that though waste minimization commonly seems an environmental welfare concentrated scheme, other sectors can be also benefitted through its implementation.

Additionally, waste minimization can also contribute to fulfill targets of other SDGs. For instance, SDG 6 aims to ensure clean water and sanitation for which wastewater and sanitation waste recycling and treatment are needed. Since industrial waste generation is responsible for significant share of wastewater disposal in the rivers and the seas, if industries take responsibility to reduce the amount of produced waste, amount of produced wastewater, contaminated by several chemical and biological hazards, will also be reduced (Vigneswaran et al., 1999). Besides, as composting toilets are often regarded as a sustainable means to inhibit exposure to sanitation waste, modifying composting strategy based on geographic and climate variables, can impact attainment of SDG 7's target positively (Anand and Apul, 2014). Here, composting is another important strategy under waste minimization. Moreover, SDG 13 suggests adopting climate friendly actions for which waste minimization initiatives can be useful. It is because several waste management strategies like incineration, landfills, etc. and untreated waste disposal are found to be responsible for carbon dioxide and other green house gases (GHGs) emission; thus there are chances that if amount of produced waste is minimized, contribution of waste to climate change will also be minimized (Schott and Andersson, 2015; Bogner et al., 2008). This is how waste minimization can be used as a climate action in achieving SDG 13.

Waste minimization and waste source reduction strategies are needed to protect our environment and mankind. Its significance can be truly understood while all of its multiple benefits are considered from a broader angle. Otherwise, benefits that waste minimization strategies can bring about, might not convince stakeholders from different sectors to implement waste minimization strategies in their respective field.

Case Study 1: Finelite Inc.'s Effort to Reduce Waste

The story of Finelite Inc. is an inspiring story of a Union City manufacturer who could make profit in an environmentally sustainable and innovation way aiming to reduce their waste generation. Primarily, they introduced a recycling program in 2006 which aimed to minimize the company's landfill contribution by reusing and recycling materials like e-waste, wood products, metals, plastics, and so on. They have also examined details of their supply chain mechanism and operation to find out every possible ways through which their net waste contribution to the society will be minimized. For instance, to minimize transportation-emitted gases, they consolidated their activities in the nearby local areas without using fueled vehicles. They replaced many of the chemicals used in industrial processes to ensure sustainability. Even while packaging their goods for shipment, unlike many industries, they started using crinkle papers and paper tapes instead of bubble wraps and staples respectively. They also took initiatives to modify all their machineries using more sustainable products. Thus with their kind efforts, they tried in every possible ways to reduce their waste generation. They have adopted many more strategies in every step of the design, manufacturing, operations, and supply chain. Adopting all such initiatives surprisingly reduced their waste from an annual waste production of 110–18 tons—an

84% reduction in waste generation within an interval of 6 years. In addition to it, they could save 17% of their packing cost annually. After a payback period below 1 year, they could save over $ 10,000 per year as a co-benefit of their waste minimization initiatives (Finelite, 2015). This indicates all such initiatives not only reduced their environmental footprint but also such initiatives brought about economic benefit of the company.

What makes the success story of the company more appropriate in this context is that most of their sustainable solutions were focused on pregeneration stage of waste. It denotes, instead of installing expensive and complex waste treatment and management technologies, their focus was more concentrated in minimizing waste at a source level, which could make it easier for them to shorten the payback period and gain more economic benefits. Creativity and detailed analysis in designing the sustainable solutions was also another crucial factor behind their success. As it been mentioned at the beginning of the chapter, prevention is better than cure, the story of Finelite can be regarded as a very suitable example in this regard. Realizing their environmental commitment of waste reduction, Finelite have been awarded StopWaste Partnership Business Efficiency Award of 2011 and 2nd Annual Reusable Packaging Association's Excellence in Reusable Packaging Award to encourage other manufacturers' contribution in solving the waste issue at source level.

References

Aadal, H., Rad, K. G., Fard, A. B., Sabet, P. G. P., & Harirchian, E. (2013). Implementing 3R concept in construction waste management at construction site. *Journal of Applied Environmental and Biological Science, 3*(10), 160–166.

Abdelmalek, F.T., & Rutsch, M.J. (1997). *U.S. Patent No. 5,642,630.* Washington, DC: U.S. Patent and Trademark Office.

AbdManaf, L., Samah, M.A.A., & Zukki, N. I. M. (2009). Municipal solid waste management in Malaysia: Practices and challenges. *Waste Management, 29*(11), 2902–2906.

Abdulla, F., Qdais, H.A., & Rabi, A. (2008). Site investigation on medical waste management practices in northern Jordan. *Waste Management, 28*(2), 450–458.

Agamuthu, P. (2009). Challenges and opportunities in agro-waste management: An Asian perspective. In Inaugural meeting of first regional 3R forum in Asia, 11–12 Nov 2009, Tokyo, Japan, University of Malaysia.

Ahmann, D., Roberts, A. L., Krumholz, L. R., & Morel, F. M. (1994). Microbe grows by reducing arsenic. *Nature, 371*(6500), 750–1750.

Al Yaqout, A. F. (2003). Assessment and analysis of industrial liquid waste and sludge disposal at unlined landfill sites in arid climate. *Waste Management, 23*(9), 817–824.

AlQattan, N., Acheampong, M., Jaward, F. M., Ertem, F. C., Vijayakumar, N., & Bello, T. (2018). Reviewing the potential of Waste-to-Energy (WTE) technologies for Sustainable Development Goal (SDG) numbers seven and eleven. *Renewable Energy Focus, 27*, 97–110.

Anand, C. K., & Apul, D. S. (2014). Composting toilets as a sustainable alternative to urban sanitation–A review. *Waste Management, 34*(2), 329–343.

Arena, U., Mastellone, M. L., & Perugini, F. (2003). The environmental performance of alternative solid waste management options: a life cycle assessment study. *Chemical Engineering Journal, 96*(1–3), 207–222.

Athanasiou, M., Makrynos, G., & Dounias, G. (2010). Respiratory health of municipal solid waste workers. *Occupational Medicine, 60*(8), 618–623.

Aydemir-Karadag, A. (2018). A profit-oriented mathematical model for hazardous waste locating-routing problem. *Journal of Cleaner Production, 202*, 213–225.

Babu, B. R., Bhanu, S. U., & Meera, K. S. (2009). Waste minimization in electroplating industries: a review. *Journal of Environmental Science and Health Part C, 27*(3), 155–177.

Bahor, B., Van Brunt, M., Stovall, J., & Blue, K. (2009). Integrated waste management as a climate change stabilization wedge. *Waste Management & Research, 27*(9), 839–849.

Blengini, G. A., Busto, M., Fantoni, M., & Fino, D. (2012). Eco-efficient waste glass recycling: Integrated waste management and green product development through LCA. *Waste Management, 32*(5), 1000–1008.

Bogner, J., Pipatti, R., Hashimoto, S., Diaz, C., Mareckova, K., Diaz, L., et al. (2008). Mitigation of global greenhouse gas emissions from waste: conclusions and strategies from the Intergovernmental Panel on Climate Change (IPCC) Fourth Assessment Report. Working Group III (Mitigation). *Waste Management & Research, 26*(1), 11–32.

Borai, E. H., Harjula, R., & Paajanen, A. (2009). Efficient removal of cesium from low-level radioactive liquid waste using natural and impregnated zeolite minerals. *Journal of Hazardous Materials, 172*(1), 416–422.

Ceylan, A., Ertem, M., Ilcin, E., & Ozekinci, T. U. N. C. E. R. (2003). A special risk group for hepatitis E infection: Turkish agricultural workers who use untreated waste water for irrigation. *Epidemiology & Infection, 131*(1), 753–756.

Chaaban, M. A. (2001). Hazardous waste source reduction in materials and processing technologies. *Journal of Materials Processing Technology, 119*(1–3), 336–343.

Chen, X., Geng, Y., & Fujita, T. (2010). An overview of municipal solid waste management in China. *Waste Management, 30*(4), 716–724.

Cheremisinoff, N. P. (2003). *Handbook of solid waste management and waste minimization technologies*. Butterworth-Heinemann.

Chi, B., Lu, W., Ye, M., Bao, Z., & Zhang, X. (2020). Construction waste minimization in green building: A comparative analysis of LEED-NC 2009 certified projects in the US and China. *Journal of Cleaner Production*, 120749.

Christensen, T. (Ed.). (2011). *Solid waste technology and management*. John Wiley & Sons.

Clark, J. H. (2012). *Chemistry of waste minimization*. Springer Science & Business Media.

Cole, C., Osmani, M., Quddus, M., Wheatley, A., & Kay, K. (2014). Towards a zero waste strategy for an English local authority. *Resources, Conservation and Recycling, 89*, 64–75.

Dhokhikah, Y., & Trihadiningrum, Y. (2012). Solid waste management in Asian developing countries: challenges and opportunities. *Journal of Applied Environmental and Biological Sciences, 2*(7), 329–335.

Domene, X., Ramírez, W., Mattana, S., Alcaniz, J. M., & Andrés, P. (2008). Ecological risk assessment of organic waste amendments using the species sensitivity distribution from a soil organisms test battery. *Environmental Pollution, 155*(2), 227–236.

Dunn, J.L., & Guerra, L.E. (2005). *U.S. Patent No. 6,893,425*. Washington, DC: U.S. Patent and Trademark Office.

EPA (2020). *Defining hazardous waste: listed, characteristic and mixed radiological wastes*. epa.gov.

Feng, G. L., Letey, J., Chang, A. C., & Mathews, M. C. (2005). Simulating dairy liquid waste management options as a nitrogen source for crops. *Agriculture, Ecosystems & Environment, 110*(3–4), 219–229.

Finelite (2015). Finelite's environmental commitment: green operations. *Finelite.com.*

Gentil, E., Clavreul, J., & Christensen, T. H. (2009). Global warming factor of municipal solid waste management in Europe. *Waste Management & Research, 27*(9), 850–860.

Giugliano, M., Cernuschi, S., Grosso, M., & Rigamonti, L. (2011). Material and energy recovery in integrated waste management systems. An evaluation based on life cycle assessment. *Waste Management, 31*(9–10), 2092–2101.

Giusti, L. (2009). A review of waste management practices and their impact on human health. *Waste Management, 29*(8), 2227–2239.

Goel, P. K. (2006). *Water pollution: causes, effects and control.* New Age International.

Graham, P. (2000). Building education for the next industrial revolution: teaching and learning environmental literacy for the building professions. *Construction Management & Economics, 18*(8), 917–925.

Grant, K., Goldizen, F. C., Sly, P. D., Brune, M. N., Neira, M., van den Berg, M., et al. (2013). Health consequences of exposure to e-waste: a systematic review. *The Lancet Global Health, 1*(6), e350–e361.

Guerrero, L. A., Maas, G., & Hogland, W. (2013). Solid waste management challenges for cities in developing countries. *Waste Management, 33*(1), 220–232.

Harremoës, P. (1997). Integrated water and waste management. *Water Science and Technology, 35*(9), 11–20.

Heidrich, O., Harvey, J., & Tollin, N. (2009). Stakeholder analysis for industrial waste management systems. *Waste Management, 29*(2), 965–973.

Hoornweg, D., & Bhada-Tata, P. (2012). *What a waste: a global review of solid waste management.* Urban development series; Knowledge paper no. 15. World Bank, Washington, DC.

Hossain, M. S., Santhanam, A., Norulaini, N. N., & Omar, A. M. (2011). Clinical solid waste management practices and its impact on human health and environment–A review. *Waste Management, 31*(4), 754–766.

Huang, B., Wang, X., Kua, H., Geng, Y., Bleischwitz, R., & Ren, J. (2018). Construction and demolition waste management in China through the 3R principle. *Resources, Conservation and Recycling, 129*, 36–44.

Hung, Y. T., Wang, L. K., & Shammas, N. K. (Eds.). (2012). *Handbook of environment and waste management: air and water pollution control* (Vol. 1). World Scientific.

Jamasb, T., & Nepal, R. (2010). Issues and options in waste management: A social cost–benefit analysis of waste-to-energy in the UK. *Resources, Conservation and Recycling, 54*(12), 1341–1352.

Javaheri, H., Nasrabadi, T., Jafarian, M. H., Rowshan, G. R., & Khoshnam, H. (2006). Site selection of municipal solid waste landfills using analytical hierarchy process method in a geographical information technology environment in Giroft. *Iranian Journal of Environmental Health Science & Engineering, 3*(3), 177–184.

Kazi, T. G., Jamali, M. K., Kazi, G. H., Arain, M. B., Afridi, H. I., & Siddiqui, A. (2005). Evaluating the mobility of toxic metals in untreated industrial wastewater sludge using a BCR sequential extraction procedure and a leaching test. *Analytical and Bioanalytical Chemistry, 383*(2), 297–304.

Khursheed, T., & Singh, B. K. (2017). Road construction from solid waste in Jaipur. *International Journal for Technological Research in Engineering, 4*(9), 1711–1714.

Lober, D. J. (1996). Municipal solid waste policy and public participation in household source reduction. *Waste Management & Research, 14*(2), 125–143.

López-González, J. A., del Carmen Vargas-García, M., López, M. J., Suárez-Estrella, F., del Mar Jurado, M., & Moreno, J. (2015). Biodiversity and succession of mycobiota associated to agricultural lignocellulosic waste-based composting. *Bioresource Technology, 187*, 305–313.

Mahat, H., Yusri, M. S., & Ngah, C. (2016). 3R practices among more preschool pupils through the environmental education curriculum. *SHS Web of Conferences, 23.* EDP Sciences.

Malviya, R., & Chaudhary, R. (2006). Factors affecting hazardous waste solidification/stabilization: A review. *Journal of Hazardous Materials, 137*(1), 267–276.
Martuzzi, M., Mitis, F., & Forastiere, F. (2010). Inequalities, inequities, environmental justice in waste management and health. *European Journal of Public Health, 20*(1), 21–26.
McDougall, F. R., White, P. R., Franke, M., & Hindle, P. (2008). *Integrated solid waste management: a life cycle inventory*. John Wiley & Sons.
Memon, M. A. (2010). Integrated solid waste management based on the 3R approach. *Journal of Material Cycles and Waste Management, 12*(1), 30–40.
Menikpura, S. N. M., Sang-Arun, J., & Bengtsson, M. (2013). Integrated solid waste management: an approach for enhancing climate co-benefits through resource recovery. *Journal of Cleaner Production, 58*, 34–42.
NA, A. B., Fauzi, M. F., Muin, A., Saati, S. A., Shafie, F. A., Sahhir, K., et al. (2020). Towards zero waste: technologies and minimisation strategies. *MAEH Journal of Environmental Health (Short Communication), 1*(1), 4–7.
Nabavi-Pelesaraei, A., Bayat, R., Hosseinzadeh-Bandbafha, H., Afrasyabi, H., & Chau, K. W. (2017). Modeling of energy consumption and environmental life cycle assessment for incineration and landfill systems of municipal solid waste management-A case study in Tehran Metropolis of Iran. *Journal of Cleaner Production, 148*, 427–440.
Nguyen, T. T. P., Zhu, D., & Le, N. P. (2015). Factors influencing waste separation intention of residential households in a developing country: Evidence from Hanoi, Vietnam. *Habitat International, 48*, 169–176.
Obi, F. O., Ugwuishiwu, B. O., & Nwakaire, J. N. (2016). Agricultural waste concept, generation, utilization and management. *Nigerian Journal of Technology, 35*(4), 957–964.
Pires, A., Martinho, G., & Chang, N. B. (2011). Solid waste management in European countries: A review of systems analysis techniques. *Journal of Environmental Management, 92*(4), 1033–1050.
Porter, R. C. (2010). *The economics of waste*. Routledge.
Ramachandra, T.V., & Varghese, S. (2003). Exploring possibilities of achieving sustainability in solid waste management. *Indian Journal of Environmental Health, 45*(4), 255–264.
Reyes, H., Guiscafre, H., Muñoz, O., Perez-Cuevas, R., Martinez, H., & Gutierrez, G. (1997). Antibiotic noncompliance and waste in upper respiratory infections and acute diarrhea. *Journal of Clinical Epidemiology, 50*(11), 1297–1304.
Rumph, R. M. (2001). *U.S. Patent No. 6,333,446*. Washington, DC: U.S. Patent and Trademark Office.
Rushton, L. (2003). Health hazards and waste management. *British Medical Bulletin, 68*(1), 183–197.
Schott, A. B. S., & Andersson, T. (2015). Food waste minimization from a life-cycle perspective. *Journal of Environmental Management, 147*, 219–226.
Seadon, J. K. (2006). Integrated waste management–looking beyond the solid waste horizon. *Waste Management, 26*(12), 1327–1336. https://doi.org/10.1016/j.wasman.2006.04.009.
Seadon, J. K. (2010). *Integrated waste management* (Doctoral dissertation, University of Auckland).
Sharholy, M., Ahmad, K., Mahmood, G., & Trivedi, R. C. (2008). Municipal solid waste management in Indian cities–A review. *Waste Management, 28*(2), 459–467.
Simon, E., Harangi, S., Baranyai, E., Braun, M., Fábián, I., Mizser, S., et al. (2016). Distribution of toxic elements between biotic and abiotic components of terrestrial ecosystem along an urbanization gradient: Soil, leaf litter and ground beetles. *Ecological Indicators, 60*, 258–264.
Song, Q., Li, J., & Zeng, X. (2015). Minimizing the increasing solid waste through zero waste strategy. *Journal of Cleaner Production, 104*, 199–210.
Soukopová, J., & Malý, I. (2013). Competitive environment in waste management and its impact on municipal expenditures. *Acta Universitatis Agriculturae et silviculturae Mendelianae Brunensis, 61*(4), 1113–1119.

Tangwanichagapong, S., Nitivattananon, V., Mohanty, B., & Visvanathan, C. (2017). Greening of a campus through waste management initiatives. *International Journal of Sustainability in Higher Education, 18*(2), 203–217.

Texier, D., Degnan, P., Loutre, M. F., Lemaitre, G., Paillard, D., & Thorne, M. (2003). Modelling sequential BIOsphere systems under CLIMate change for radioactive waste disposal. Project BIOCLIM. In *Proceeding of the tenth international high level radioactive waste management conference* (pp. 202–212). Las Vegas, USA.

Titus, C. H., & Surma, J. E. (2001). *U.S. Patent No. 6,215,678.* Washington, DC: U.S. Patent and Trademark Office.

Tóth, A. J. (2015). *Liquid waste treatment with physicochemical tools for environmental protection.* PhD Thesis, Department of Chemical and Environmental Process Engineering, Budapest.

Van Berkel, R. (2004). Assessment of the impact of the DESIRE project on the uptake of waste minimization in small scale industries in India (1993-1997). *Journal of Cleaner Production, 12*(3), 269–281.

Vasanthi, P., Kaliappan, S., & Srinivasaraghavan, R. (2008). Impact of poor solid waste management on ground water. *Environmental Monitoring and Assessment, 143*(1–3), 227–238.

Vigneswaran, S., Jegatheesan, V., & Visvanathan, C. (1999). Industrial waste minimization initiatives in Thailand: concepts, examples and pilot scale trials. *Journal of Cleaner Production,* 7(1), 43–47.

Weng, Y. C., & Fujiwara, T. (2011). Examining the effectiveness of municipal solid waste management systems: an integrated cost–benefit analysis perspective with a financial cost modeling in Taiwan. *Waste Management, 31*(6), 1393–1406.

Wilson, D. C., Velis, C., & Cheeseman, C. (2006). Role of informal sector recycling in waste management in developing countries. *Habitat International, 30*(4), 797–808.

World Bank (2019). Understanding poverty urban development: solid waste management. *Worldbank.org.*

Zaman, A. U. (2015). A comprehensive review of the development of zero waste management: lessons learned and guidelines. *Journal of Cleaner Production, 91*, 12–25.

Zhang, D. Q., Tan, S. K., & Gersberg, R. M. (2010). Municipal solid waste management in China: status, problems and challenges. *Journal of Environmental Management, 91*(8), 1623–1633.

Zorpas, A. A., & Lasaridi, K. (2013). Measuring waste prevention. *Waste Management, 33*(5), 1047–1056.

Further reading

Carden, J. L., Jr. (1985). Hazardous waste management. *Introduction to environmental health.* New York: Springer-Verlag Inc..

Morris, G. E., & Holthausen, D. M., Jr. (1994). The economics of household solid waste generation and disposal. *Journal of Environmental Economics and Management, 26*(3), 215–234.

Osmani, M. (2012). Construction waste minimization in the UK: current pressures for change and approaches. *Procedia-Social and Behavioral Sciences, 40*, 37–40.

CHAPTER 2

Source reduction, waste minimization, and cleaner technologies

2.1 Introduction

Most of the source reduction and waste minimization strategies are technology-mediated and these technologies are known as cleaner technologies. Such cleaner technologies refer to technologies which are efficient enough to minimize waste production and are ecofriendly in terms of their environmental impacts. As discussed in the previous chapter, waste minimization is a broad and complex issue. Besides, source-level minimization of waste is even more complex considering suitable technological unavailability and industrial process complexity. Besides, in low-income countries where there is scarcity of waste management initiatives, individuals and industries of such countries are often more unwilling to invest on source-level waste minimization. However, the benefits brought by cleaner technologies are undeniable. The variety of cleaner technologies offers the users to decide on technologies as per their demands and corresponding context.

Adoption of cleaner technologies for source reduction and waste minimization is beneficial for several purposes. First, source-level waste minimization technologies offer destruction of waste before its exposure. This is why produced waste cannot cause any harm to the environment and public health. Thus, source-level waste reduction minimizes waste-borne hazards largely. Second, in case of source-level waste minimization, individuals get the opportunity to minimize their produced waste. Thus, the waste producers are assigned the responsibilities of waste minimization. This is how as everyone is responsible for reducing their personally generated waste under a waste minimization scheme, it becomes easier to be tackled successfully. Third, waste source reduction initiatives lower waste transportation cost largely. As waste transportation and collection is often more harmful and more expensive than process associated with many minimization strategies, treating the waste at source would reduce cost and hazards associated with transportation and collection (Zsigraiová et al., 2009). Besides, cleaner

Source Reduction and Waste Minimization
http://dx.doi.org/10.1016/B978-0-12-824320-6.00002-2

technologies adopted for waste source reduction have several co-benefits and can play significant role in achieving sustainable development goals (SDGs) for which different group of stakeholders should also be interested in adopting these technologies. Thus, cleaner waste source reduction technologies are needed to maintain long-term sustainability in waste sector.

This chapter explores several cleaner technologies and cleaner initiatives which are effective for waste minimization both in large and small scale. At first, it describes some of its unique characteristics and benefits. Then gradually this chapter discusses on several different waste minimization technologies and approaches like recycling, composting, biological treatment, chemical treatment, thermal treatment, filtration, energy source modification, and so on. Finally, it discusses on the role of green chemistry principles for successful waste minimization. It also includes suitable case studies to understand real-life implementation of cleaner technologies for source reduction and waste minimization.

2.2 Recycling

Recycling is the resource circulation process through which waste materials are converted into useful goods. It is an integral component of 3R approach. To minimize waste materials at source level, recycling of goods is a primary need. Among the major benefits of recycling, some of the most notable benefits are resource recovery, minimizing expenditure, less environmental exposure to waste materials, minimization of waste-borne health hazards, and so on. Considering all the benefits of recycling strategy, it is urgent to introduce more recycling strategies from small to large sectors. Recycling itself is a broad and complex process, as recycling strategies vary largely among different sectors. There are several criteria that should be considered before recycling a product. First, recycling process should be affordable and cost-effective (Olivieri et al., 2014). Second, recycling strategies should not be harmful for our environment and health that denotes these strategies should fully harmless (Brissaud, 2008). Third, recycling strategies should be socially acceptable based on social decision making (Wibowo and Deng, 2015). It denotes that recycling process should consider the social and ethical values of a society before implementing. For example, sanitary wastes are often used in agricultural fields after composting. But if a society denies to adopt this strategy due to their social or religious ethics, they should not impose this strategy forcefully. Thus, there are many more things which should be considered before implementing a successful recycling strategy (Fig. 2.1).

Figure 2.1 ***Closed loop to implement sustainable recycling.***

The application and variance strategy depends largely in different sectors. For instance, at an industrial level, through recycling, a wide variety of products are reused that includes steel slag, bricks, glass or ceramic products, and so on (Pelino et al., 1997; Li et al., 2009; Andreola et al., 2005). In an informal setting, recycling is often adopted by people of developing countries to improve their livelihood utilizing minimum resources. For instance, the lower-income communities often build their household using grasses and disposed woods or cloths (Wilson et al., 2006). Besides, solid wastes such as plastics, ceramic, or glasses are often recycled to produce new products (Troschinetz and Mihelcic, 2009). Other types of waste materials like oyster shells are often used to control eutrophication in coastal regions of southeast Korea (Kwon et al., 2004). Thus, in various sectors, the implication of recycling varies largely.

2.3 Composting

Composting is an age-old traditional waste minimization strategy. Composting denotes an aerobic method of decomposing organic solid waste. It is a major type of recycling organic waste to produce useful fertilizers. Composting produces humus as an end product which is beneficial for plant growth and effective to minimize organic waste at small or large scale. Composting process can be applicable for a wide variety of waste components like solid and liquid waste. Based on the products which are being composted, suitable ingredients are designed to enhance composting procedure (Misra et al., 2003). Usually animal feces or plants are used as composting ingredients. In this process, moisture content of the waste materials are

Figure 2.2 ***Basic steps of composting process.***

reduced drastically and in presence of oxygen, the organic parts of the waste materials are decomposed at certain temperature (Richard et al., 2002). Usually, warm temperature of tropical regions is more suitable for composting comparing with the cooler regions (Liang et al., 2003). Some basic steps of composting strategy have been illustrated in Fig. 2.2.

Composting process occurs in a series of chemical reactions. Based on the primary sources used in composting, the reaction steps and mechanisms vary. Some of such major biochemical reactions involved with composting process have been illustrated in Fig. 2.3 (Peigné and Girardin, 2004).

In brief, as an end product of the process, carbon rich humus is produced which is rich in plant nutrients. Composting process can be of various types considering several criteria. Some of the common ones are windrow composting, tunnel composting, in-vessel composting, composting toilet, and so on. Currently it is a very well-known process adopted by many people worldwide. However, it is not fully ecofriendly as composting plants often release methane and other harmful gases (Jäckel et al., 2005). But comparing with many other waste minimization strategies, it is quite effective in terms of its affordability and effectivity.

2.4 Biological treatment

Several biological processes have been found useful for source-level waste minimization. These techniques are applicable for both industrial and household level because of less risks associated with such processes (Deleris

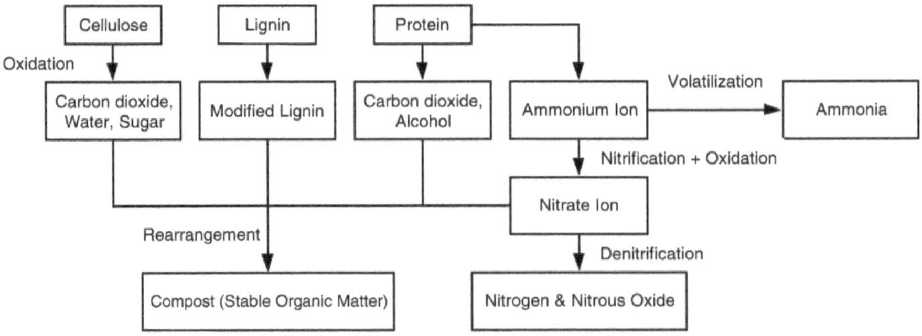

Figure 2.3 ***Major biochemical reactions occurred in composting process.***

et al., 2002). The efficiency of these techniques depends on several parameters of the used microorganisms and chemicals. Based on the characteristics of the microorganisms and chemicals used in this process, specific hazardous chemicals are reduced from the waste materials (Silva et al., 2002; Paul and Liu, 2012). It denotes that these processes are not universal to reduce all different types of waste materials. However, some of the efficient biological processes of waste minimization are discussed in the following section.

2.4.1 Anaerobic digestion

Anaerobic digestion is the process through which microorganisms break down the biodegradable compounds in anaerobic environment. Among its multiple implications in various sectors of waste management, this process is widely used for wastewater and sludge treatment (Appels et al., 2008; Deleris et al., 2002). This process is also found useful to minimize amount of produced sludge that can be considered as a source-level sludge minimization technique (Wei et al., 2003). Besides, anaerobic digestion is also applicable to minimize waste produced in food industries, as food components are mostly biodegradable by anaerobic microbes. Anaerobic digestion is also favorable to treat organic solid waste but in such regards, choosing suitable microbes is necessary (Khalid et al., 2011). In the industries, anaerobic digestion is often introduced by installing UASB (Up-flow Anaerobic Sludge Blanket) or facultative lagoons. Facultative lagoons are historically used as anaerobic digesters due to their technological simplicity and efficiency as anaerobic digester. These lagoons are very efficient in minimizing complex sludge and liquid organic waste using a less expensive workflow (Hamilton, 2004). Implementation of anaerobic digester in recent days is associated with UASB technology where industries form a sludge blanket that is degraded by anaerobic microorganisms. Besides, it is often regarded as a very environment friendly technology, as its implementation in the industries found significant economic benefit from production of biogas and bio-oil along with huge reduction of waste-borne carbon dioxide (Chinnaraj and Rao, 2006). Apart from industrial implementation, cities like San Francisco could successfully introduce anaerobic digestion at household level food waste treatment (Zhang et al., 2007). Thus, anaerobic digestion, being a popular biological treatment methodology can help industries in minimizing several different types of waste at a source-level.

Here, Fig. 2.4 illustrates an overview on basic biochemical reactions of anaerobic digestion.

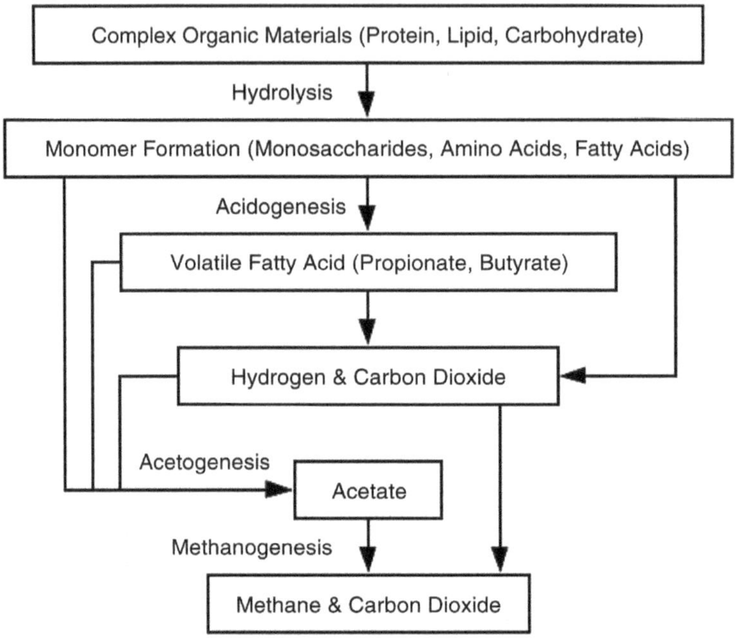

Figure 2.4 ***Summary of biochemical reactions for anaerobic digestion.***

In different reaction mechanisms, anaerobic digestion breaks down larger biochemical compounds into smaller compounds. The microorganisms used in this process regulate secretion of suitable enzymes to assist the digestion process. The pace and effectivity of the process is also maintained by suitable carbon-nitrogen ratio of the used waste materials.

2.4.2 Enzyme-based waste minimization

Enzymes are often used to reduce waste of various types. In such regards, either enzyme solutions are directly mixed with waste materials or enzyme producing bacteria interact with waste materials. This process is very specific, as any specific enzyme cannot universally degrade all different types of waste materials. For instance, phenol degradation from waste requires intervention of any enzyme from a specific group of enzyme that is oxidoreductase (Narayan and Agrawal, 2012). Using such enzymes, phenol can be removed from a mixture of waste components in a low retention time. The efficiency of the process is quite high for which such enzyme can degrade phenol from waste even if the phenolic concentration is quite low (Narayan and Agrawal, 2012). In this regard, any oxidoreductase producing bacteria can also be used instead of using oxidoreductase enzyme directly. Besides, though many enzymes can remove a certain waste component from a chunk of waste, the kinetics might be different for each of the enzymes (Melikoglu et al., 2013). It denotes that the sensitivity and retention time will be different for different enzymes. Based on the corresponding kinetics factors, the suitable enzyme is chosen to reduce a certain type of chemical. This process can be implemented both for solid and liquid waste minimization. To treat solid wastes using enzyme-based method, suitable enzymatic assay is needed to be prepared so that the compounds of interests can be isolated from the solid waste (Rodriguez et al., 2005). This process is an energy efficient and environment friendly procedure. The major challenge associated with this process is to find out the suitable enzyme for waste minimization.

In this process, enzymes catalyze waste minimization mechanisms in different ways such as biocatalysis, biotransformation, and biodegradation (Parales et al., 2002). Using these processes, enzymatic influence either degrades or transforms waste materials or catalyzes waste minimization processes. All these different ways of waste minimization should be interconnected for successful waste reduction. Otherwise, incomplete implementation of any of these steps will cause failure to proper waste minimization, as hazards or risk caused by exposure to waste materials will be still prevalent (Fig. 2.5).

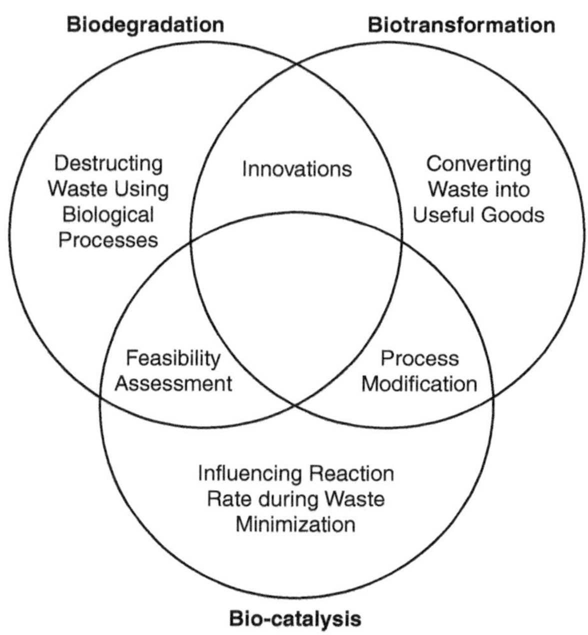

Figure 2.5 ***Ways of enzyme-mediated waste minimization presented in Venn diagram.***

2.4.3 Membrane-based system

Membrane-based system for waste minimization is a biological component mediated ultrafiltration process. The membrane can be prepared using different microorganisms or biological bodies. Contextualizing membrane components used in filtration process solid, liquid, and gaseous waste materials can be treated. Based on the components of the membrane, the filtration potential as well as waste minimization potential is determined (Kumar et al., 2008). This process is mostly used in minimizing heavy metals from waste materials and for wastewater treatment. In this regard, liquid membranes are often used that differentiate heavy metals from waste components in terms of their electron configuration and size (Bhatluri et al., 2015). Other types of organic waste components can also be separated using suitable membranes from industrial effluent. This process requires low capital cost for implementation and it can be regarded as quite energy efficient and ecofriendly (Raza et al., 2019). Designing suitable membrane based on the industrial need is the only complex and time consuming step of this process.

Membrane-based waste minimization technology helps to minimize a variety of waste materials in various ways. It can minimize solid, liquid, and gaseous waste materials using different membranes and multiple isolation strategies. For example, in case of liquid waste, membranes having pores of different sizes and adhesive nature, molecules of different hazardous waste materials are filtered. Another advanced membrane-based waste minimization technique, membrane condenser technique helps to extract pure water from evaporated waste. In this process of resource recovery, a semipermeable membrane is used that is selective to all the other gaseous materials transmission except vapor. Then the condensed vapor is precipitated and extracted in liquid form. Fig. 2.6 illustrates a basic structure of a membrane based waste minimization system (Macedonio et al., 2012).

Thus, different membrane-based processes contribute as waste minimization strategies offering co-benefits like resource recovery, minimized pollution, etc., along with.

2.4.4 Waste autoclave

Waste autoclave can be regarded as a biological process of waste minimization, as autoclave process is focused on the removal of biological bodies. Autoclaving clinical waste before disposal is very useful to hinder exposure to clinical waste that may cause emergence of various diseases. Even autoclaving clinical waste is a better option than incinerating clinical waste, considering the environmental and economic benefits (Hossain et al., 2012).

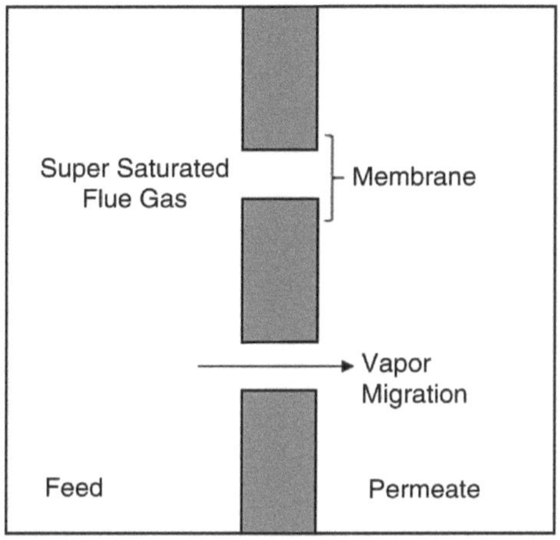

Figure 2.6 ***Schematic diagram showing membrane condenser process of waste minimization.***

In addition to it, autoclaving medical equipment is needed to sterilize the equipment and to prevent transmission of diseases from one person to other persons (Tiller and Linscott, 2004). Apart from the clinical implications of waste autoclave, it can be very relatable with laboratory waste minimization; especially in case of bacteriology and microbiology laboratories. Equipment used for multiple times require autoclave before usage. The accuracy of the experimental findings might be hampered, unless autoclaving equipment is not maintained at a regular basis (Armstrong and Reinhardt, 2010). Though it is not a direct waste minimization technology, it can be considered as a waste hazard minimization strategy at source level.

2.5 Chemical treatment

Waste minimization and source reduction can be associated with several chemical strategies that will help both to minimize amount of waste and to recover useful resources from waste materials. Chemical treatment of waste requires installation of sophisticated setting and utilizing suitable chemicals based of the types of waste generation by the industries. By integrating concepts of chemical waste minimization processes with other waste minimization strategies, newly modified waste minimization strategies are often designed (Marculescu, 2011). While designing useful chemical strategies for waste minimization, waste generation process, chemical of produced waste, and its cause-effect functional elements should be analyzed in depth (Halim and Srinivasan, 2002). Then based on the stakeholders' demand and necessity, certain chemicals found in waste frequently are minimized through waste minimization process. Electrolysis and pyrolysis are two common chemical waste minimization strategies.

2.5.1 Electrolysis

Electrolysis is the process that uses electric current to run a non-spontaneous chemical reaction. It is a commonly used reusing technique that recovers useful resources from waste materials. It is mostly used to isolate valuable metals and organic compounds from waste in electrolysis cells (Zhang and Angelidaki, 2014). Electrolysis cells are designed specifying minimization of certain waste types. In this regard, microbial electrolysis cells can be a suitable example that combines biological and chemical waste minimization techniques to ensure better yield of recovered resources (Zhang and Angelidaki, 2014). Such cells are still an emerging technology that has potential to generate hydrogen gas, biofuels, and other organic compounds recovery (Lu and Ren, 2016). In this process, a voltage supply is required to smoothly run the process ensuring energy supply. Fig. 2.7 illustrates a basic structure of such microbial electrolysis cell that can be modified to minimize waste (Kadier et al., 2016).

Innovative researches are going on to improve the efficiency of the electrolysis cells in terms of resource recovery by combining other chemicals in the cells. For instance, integration of ferrous activated persulfate oxidation with an electrolysis cell is found to be able in disrupting the protective barrier of complex waste materials present in sludge (Zhen et al., 2013).

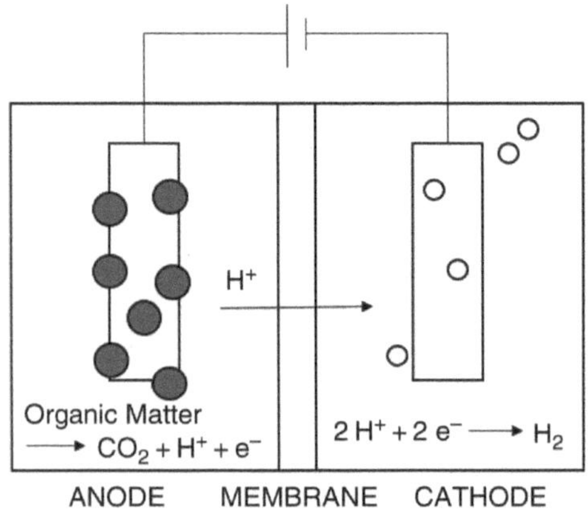

Figure 2.7 ***Schematic representation of microbial electrolysis cell used for waste minimization.***

Electrolysis can also help to treat waste leachate from both solid and liquid waste that will minimize exposure and hazard cause by waste materials (Vlyssides et al., 2001). However, considering expenses and energy efficiency, microbial electrolysis cells are more common in waste minimization plants. However, as this process also requires high energy and harmful chemicals, it is not fully ecofriendly. But the efficiency of the process is quite satisfying.

2.5.2 Pyrolysis

Pyrolysis is the thermochemical process that occurs in absence of oxygen to decompose biodegradable materials into smaller compounds at high temperature. Unlike electrolysis, pyrolysis process combines chemical and thermal waste minimization strategies. The basic difference in between pyrolysis and incineration process is that incineration occurs by oxidizing carbon containing waste while pyrolysis occurs in absence of oxygen. This is why pyrolysis process is more environment friendly than incineration, as during pyrolysis not much carbon dioxides and carbon monooxides are produced due to absence of oxygen as a reactant. Though pyrolysis process can be used for both solid and liquid waste materials, it is mostly used for solid waste minimization. It can reduce a wide range of waste materials very efficiently. In this process, mostly, carbon residues and volatile matters are recovered from organic waste (Martínez et al., 2013). In this process, minimization of solid waste, plastics, sludge, and other solid waste have been possible (Chen et al., 2014; Karayildirim et al., 2006). Pyrolysis may seem an efficient process in terms of its resource recovery potential, but implementation of this process is associated with multiple hardships including energy inefficiency, carbon emission, complex installation strategies, and so on (Van Zwieten et al., 2010). Besides, maintaining suitable pressure and chemical control is also quite tricky without experts (Bartoli et al., 2015). However, proper modifications and contextualization of the process are necessary while implementing at an industrial level. Fig. 2.8 summarizes a basic pyrolysis based waste minimization strategy where pyrolysis reactor should maintain proper reaction time, feed rate, gas-flow rate, and temperature for smooth process implementation (Guedes et al., 2018).

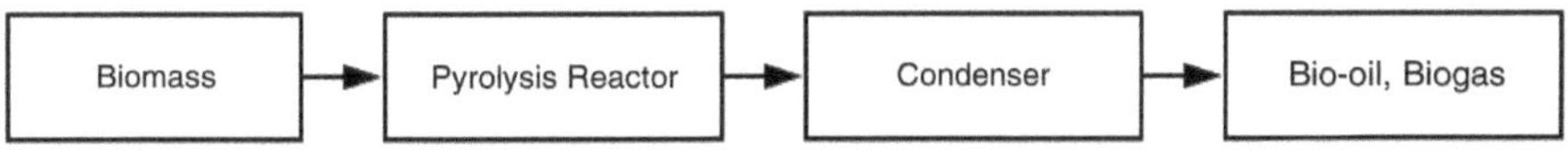

Figure 2.8 ***Waste biomass minimization and resource recovery using pyrolysis process.***

2.6 Thermal treatment

Primarily, thermal treatment of waste denotes steps through which waste materials are decomposed in presence of heat and oxygen. Thermal treatment of waste is a universally used, cost-effective means of waste treatment and minimization. As per the available facilities of a thermal treatment plant, its environmental impacts are controlled. For instance, if an incineration plant introduce carbon filter, it will have less carbon emission and if a thermal treatment plant introduce renewable energy as its energy source, it will be energy efficient. Thus, the overall benefits and negative impacts of thermal treatment initiatives largely depend of its available facilities. Among all the thermal waste treatment strategies, incineration is the most popular strategy. There are some comparatively less popular strategies like molten salt oxidation, hydrothermal carbonization, hydrothermal liquefaction (HTL), and so on.

2.6.1 Incineration

Incineration is a widely used thermal treatment mechanism for solid as well as bio-waste management. It denotes burning waste at high temperature with an aim to convert them into basic carbon compounds and heat energy. It is one of the most popular waste management technologies due to its technical simplicity and energy harvesting potential irrespective of the geographical location (Assamoi and Lawryshyn, 2012). Besides, despite having a huge construction cost, it can destroy the carcinogenic and pathogenic components found in waste materials for which incineration is regarded to have several health related co-benefits (Emek and Kara, 2007). Realizing several user-friendly characteristics of incineration technology, nowadays more and more stakeholders are being interested to install incineration facility. Based on the users' demand, incineration technology can be adopted both for large- and small-scale waste management. Large- and small-scale incineration facilities mostly vary in terms of the amount of waste to be treated and corresponding heat components (Emek and Kara, 2007). Especially, the industries often adopt small- to large-scale incineration facilities to treat their solid and organic waste. These industries prefer incineration over other waste minimization strategies as incineration is found to be useful in reducing nitrogen and sulfur oxides, two major industrially generated harmful group of chemicals, ensuring smooth combustion of waste (Nyashina et al., 2019). Besides, in densely populated municipalities where there is lack of land resource and

Figure 2.9 ***Basic workflow of incineration technology.***

technological facilities, installation of incineration facilities is quite popular in such regions (Kuo et al., 2008). As incineration facilities can often be adopted by the stakeholders who are producing the waste, it can be regarded as a prominently implemented source-level waste minimization technology (Fig. 2.9).

The mechanism, through which incineration technology functions, is not quite complex. The book *Introduction to Hazardous Waste Incineration* discusses the principles, equipment, and methodology—related to incineration technology elaborately. Historically, incineration facility didn't incorporate any extra facility other than combustion of waste at high temperature while gradually incineration technology is incorporating other facilities along with combustion. Some of such facilities include heat recovery, pollution control, standardization of parameters, and so on (Santoleri et al., 2000). Even modern incineration facilities often incorporate waste-to-energy (WtE) and energy-from-waste (EfW) concept. Additionally incineration facility can be modified with an aim to get rid of certain hazardous chemicals from the waste which requires installation of additional facilities along with traditional incineration facilities (Ioana, 2010). Thus day by day, modern incineration facilities are being redesigned based on users' preferences and needs.

Despite having several extremely beneficial outcomes of incineration facility, it is not flawless, as there are several challenges and undeniable drawbacks associated with incineration technology implementation. Incineration facility installation is quite controversial due to some of its environmental drawbacks. Most importantly, incineration is a source of gaseous pollutant emission in our environment (Ioana, 2010). It is because, in the process of waste combustion, carbon-derived gases are emitted in the environment that may contribute to climate change and different health hazards like respiratory diseases, cardiac failure, and many more health issues in later terms. This is a leading cause for which incineration technology is still not well-accepted by all the different groups of stakeholders. As a result, considering standardized parameters for such pollutants is necessary. All the existing incineration facilities should follow the regulation framework as per the country policy guidelines (Santoleri et al., 2000). To do so, many additional facilities like activated carbon filtration, carbon capture, chemical treatment, etc. are being introduced in correspondence with incineration technology (Scala, 2001). It is often a huge challenge to incorporate such additional facilities along with incineration technology due to the increased cost and higher dimensions of complexity. However, though incineration can solve co-current problems associated with waste management, considering longer terms, it cannot be regarded as a permanent and sustainable solution. This is why adoption of more efficient waste minimization technologies is needed to solve the waste issue drastically.

2.6.2 Molten salt oxidation

Molten salt oxidation is an emerging thermal waste minimization mechanism that can be used for organic waste treatment in presence of suitable salt. It is comparatively less popular than incineration and basic heat treatment. It is not widely implemented while it has potentials to be adopted in chemical industries to treat their organic waste using a very efficient mechanism (Hsu et al., 2000). This process occurs at high temperature in presence of carbonate salt by catalytic non-flame oxidation of organic waste. Physical contact in between salt bed and organic waste is an indicator of the reaction efficiency in this process (Hsu et al., 2000). Though this process causes production of harmful gases and radionuclides as by-products, it is comparatively more affordable and energy efficient than many other thermal waste treatment processes, as it can destroy 99.99% of diverse organic waste using its reactive potential (Yao et al., 2011). Industrial implementation of this process will help the industries to efficiently treat their organic waste using very limited resource. Moreover, this process has huge metal recovery potential that can be considered as a recycling strategy at industrial level (Flandinet et al., 2012). This will help the industries to reduce their waste as well as gain valuable resources. However, installation difficulty of molten salt oxidation facility is still a challenge to ensure source-level waste minimization. Therefore, it requires more modifications as per the industrial contexts. A molten salt oxidation system suitable for industrial waste minimization has been illustrated in Fig. 2.10 (Hsu et al., 2000).

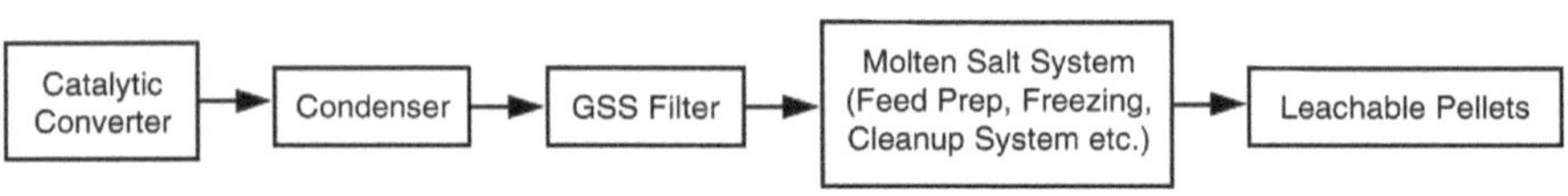

Figure 2.10 ***Integrated molten salt oxidation system for waste minimization.***

2.6.3 Hydrothermal carbonization

Hydrothermal carbonization is a novel thermal treatment based approach that can reduce waste by converting them into carbonaceous residues using steps of various chemical reactions. This process requires less temperature than all the earlier-mentioned thermal treatment mechanism for which it is a very energy efficient process. Besides, as it doesn't produce much GHGs, adoption of this process is hoped not to contribute much for climate change. This is why hydrothermal carbonization is environment friendly and quite sustainable thermal waste minimization mechanism (Lu et al., 2012). Despite being a newer waste minimization mechanism, this process is found to be a promising inclusion in waste management of energy industry, pharmaceutical industries, and for carbon sequestration (Hu et al., 2010). In this process, waste materials are heated at optimum temperature and pressure using certain plants and chemicals and then useful carbon compounds are derived within a short span of time (Lu et al., 2012). However, hydrothermal carbonization process occurs at a high temperature through a series of different types of reactions like dehydration, polymerization, and so on. Fig. 2.11 illustrates brief mechanism of hydrothermal carbonization steps at high temperature (Nicolae et al., 2020).

In the earlier-mentioned reaction mechanism, breakdown of polymers, that is, cellulose, are shown in hydrothermal carbonization process. Likewise, in case of other biodegradable waste, such steps of multiple reactions are conducted at an optimum temperature using this process. It can also be considered as a resource circulating system. The carbonaceous materials produced in this process have several industrial applications. Due to consumption of less energy and utilization of commonly found resources in this process, this process is quite affordable and beneficial for resource savings (Basso et al., 2016). Though this process was utilized in various sectors for many years, its implication in waste sector is quite new (Ramke et al., 2009). This is why its industrial application is very limited so far. But in small-scale municipal solid waste management, this process has shown auspicious success potential (Berge et al., 2011). However, its industrial implementation is not much well-established so far and there are huge scope of research and technological innovations to improvise its industrial implications.

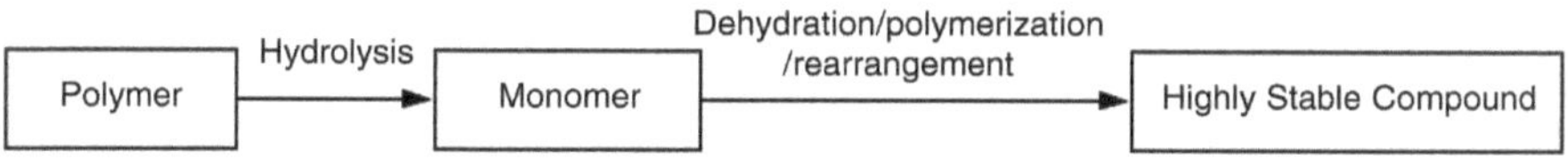

Figure 2.11 ***Hydrothermal carbonization mechanism under mild processing at 180°C–280°C temperature.***

2.6.4 Hydrothermal liquefaction (HTL)

HTL is another biopolymer waste minimization strategy that can produce value-added crude oil at the end of the process. This process incorporates thermal depolymerization reaction where waste materials are broken down into molecules of crude oil. Unlike many other thermal waste minimization techniques, HTL doesn't produce carbon containing GHGs, as all the carbons present in biopolymers are converted into monomers (Dimitriadis and Bezergianni, 2017). This process also requires several catalysts for which the temperature of thermal breakdown doesn't need to be very high comparing with incineration and other conventionally followed waste minimization approach. This is why it is quite environmental friendly process. Besides, as valuable crude oil is produced as final products of HTL, this process is also economically beneficial. In this process, biopolymers are depolymerized at optimum temperature and in the presence of catalysts that ensures high yield of crude oil. The efficiency of the process depends on the biopolymer content mostly; this is why, in case of agricultural and forest waste containing good amount of cellulose, this process has been found very efficient in terms of resource recovery and energy savings (Cao et al., 2017). Dairy sector can also be benefitted by installing this technology as a useful means to minimize dairy waste. Solid waste from municipality, food industries can also be depolymerized into crude oil using this technique (Maddi et al., 2017). Different types of waste should undergo different types of processes to produce valuable resources. For example, from the mentioned flowcharts, a comparison in between the processes involved with biomass and microalgae minimization using HTL shows that to gain hydrocarbon fuels, the adopted strategies for different waste materials are not alike. Fig. 2.12 shows how biomass can undergo HTL process and thus, it can produce hydrocarbon fuel (Biller and Ross, 2016).

Sometimes recovered resources from this process are not quite sustainable, as recovered resources like hydrocarbon fuels are often responsible for

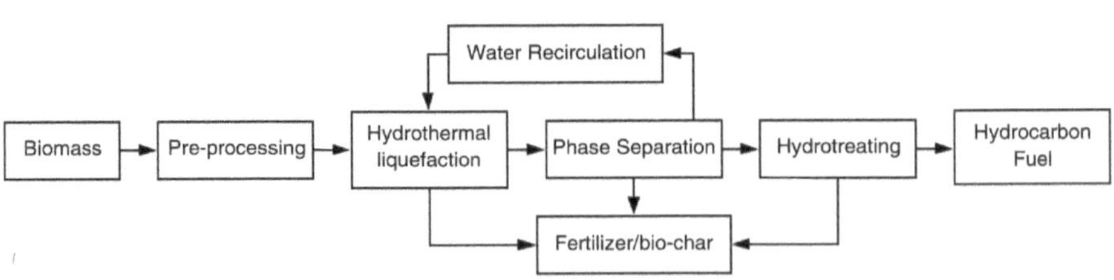

Figure 2.12 ***Steps to recover resources from biomass and using hydrothermal liquefaction process.***

high share of GHG emission. In addition to it, though HTL process has potentials in terms of waste minimization efficiently, its large scale industrial implementation with an aim to minimize waste is not implemented yet due to its failure to treat other waste except biopolymers which is a major limitation of this process.

2.6.5 Plasma gasification system

Plasma gasification is another thermal process that uses plasma to convert biodegradable waste materials into synthesis gases like carbon monoxide and hydrogen at an extremely high temperature. It is yet a very new technology that is used in waste treatment. A plasma gasification reactor is very compatible with fuel cells (Galeno et al., 2011). As per the currently available facilities under this technology, it can only treat biomass and can convert harmful compounds of biomass into comparatively less harmful compounds (Sanlisoy and Carpinlioglu, 2017). This process is also energy efficient as it can recover heat from the produced syngas. This is why plasma gasification system refers to waste to energy concept which is a very crucial waste minimization strategy. In this process of cooling and heat recovery, the pollutants are also separated from the syngas (Nielsen et al., 2015). The principle reaction in designing a model of plasma gasification reactor is as given (Achinas and Kapetanios, 2013).

$$\mathrm{CxHyOzSkNl} + \mathrm{wH_2O} + \mathrm{mO_2} + 3.76\mathrm{mN_2} \Leftrightarrow \mathrm{n1CO} + \mathrm{n2H_2} + \mathrm{n3CH_4} + \mathrm{n4H_2O} + \mathrm{n5CO_2} + \mathrm{n6N_2} + \mathrm{n7S} + \mathrm{n8H_2S}$$

Though this process produces several harmful gases, it can be considered more ecofriendly than traditional thermal treatment of waste that releases significant share of carbon dioxides and other GHGs. But yet plasma derived waste minimization is a technologically complex strategy that require further studies to be widely implemented at an affordable cost.

2.7 Filtration

Filtration is a separation strategy that isolates compounds from a mixture of substances based on its size and shape. It is a widely used liquid waste minimization strategy that includes use of various types of membrane for filtration technology. Based on the compounds of interest to be separated from the mixture and based on the membrane used in filtration, this process can be classified into classes like mechanical, biological, and chemical filtration (De Gioannis et al., 2009). In case of waste management, filtration is a very popular means to separate harmful substances. Some of the widely used filtration methodologies in case of waste minimization have been discussed herewith.

2.7.1 Biofiltration

Biofiltration is a widely used filtration technology in terms of air pollution control and liquid waste separation. In such cases, the membrane used for filtration is mostly made up of biological bodies. Some of such commonly used biofilters are anaerobic filter, vermifilters, and so on. The filtration potential of biofilters is regulated by the concentration of pollutants, pressure drop, filter bed's moisture contents, temperature, available nutrients for microorganisms, pH, and so on (De Gioannis et al., 2009). These biofilters are such strong air pollutant controller that they can even control odor caused by waste materials. And it is a very cost-effective and simple technology for odor removal (McNEVIN and Barford, 2000). Apart from treating gaseous waste, biofilters have been regarded as a quite successful strategy for liquid waste minimization by isolating organic compounds from liquid waste. In this regard, the microorganisms used as biofilters are chosen such that they can isolate and treat such organic waste into comparatively safer and cleaner compounds (Tomar and Suthar, 2011; Kumar et al., 2011). Among the widely used biofilters, vermifilters and anaerobic biofilters (made up of anaerobic microorganisms) are the most common ones. Such biofilters are also crucial for membrane based chemical treatment of waste. Fig. 2.13 represents step by step mechanism of vermifiltration process (Zhao et al., 2010).

2.7.2 Carbon filtering

Carbon filtering is the process of carbon compound isolation using chemical filters as isolation units. Such filters are mostly used for liquid and gaseous waste

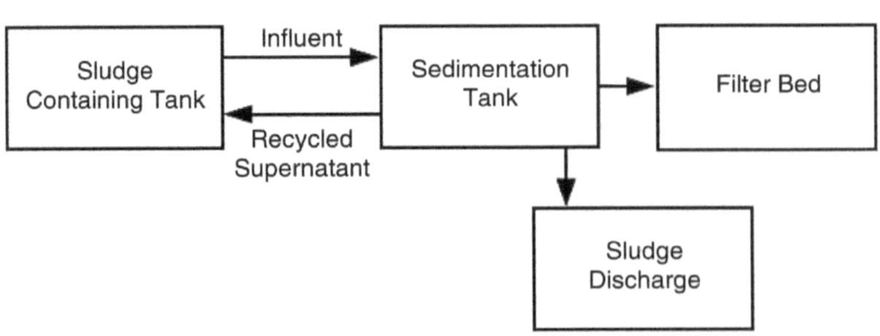

Figure 2.13 ***Schematic diagram showing vermifiltration (one of the most important biofiltration mechanism) to reduce sludge waste.***

minimization. Besides, there are evidences of processed solid waste minimization using carbon filters. Such carbon filters are often used in industries to the minimization of their pollution footprint. For instance, dye and textile industries often use electrocoagulation-biological activated carbon filtering reactor to separate dyes from wastewater for wastewater disposal (Meiqiang, 2003). Use of limestones and carbon sheets to remove ammonia nitrogen from waste leaching is also another common carbon filtering process and this process can also remove other harmful heavy metals (Aziz et al., 2004). Apart from such traditionally used filtration techniques, other advanced filtration methods are often adopted. For instance, use of carbon nanotubes to produce carbon filters is yet an emerging nanotube technology and often such nanotubes are synthesized from naturally available carbon sources (Suriani et al., 2010). Thus, carbon filters are often used as waste minimization technology to isolate and treat harmful carbon compounds. Set-up for such filters can be of various types. Such an illustration has been illustrated in Fig. 2.14 (Hasan et al., 2017).

2.7.3 Ultrafiltration

Ultrafiltration is a specialized membrane filtration technology that enhances pressure-mediated suspension of pathogenic and solid waste from waste mixture. The product produced after ultrafiltration is highly pure and free of any pathogenic waste. Most of the characteristics of ultrafiltration are quite similar to membrane-based filtration as ultrafiltration strategy is aligned with semipermeable membrane utilization. The membranes used for ultrafiltration process can be of various types. For instance, micellar enhanced membrane

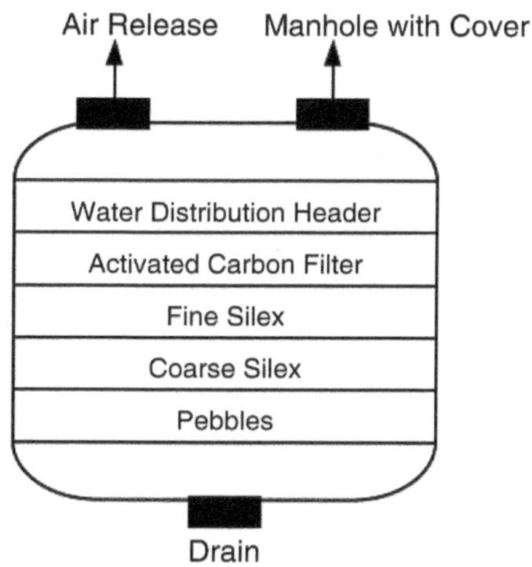

Figure 2.14 ***Set-up for activated carbon filtration process.***

can filtrate complex chemicals using pressure-driven ultrafiltration process (Bielska and Szymanowski, 2006). Both micellar and surfactant-based filters can remove heavy metals from waste using ultrafiltration strategies (Tortora et al., 2016; Huang et al., 1994). Besides, artificially produced chemical membranes are also found useful to isolate a wide range of pollutants using ultrafiltration strategies (Huang et al., 2015). Thus, as per the users demand and necessity, ultrafiltration membranes are designed to separate specific types of pollutants from waste materials using a pressure-driven process.

Fig. 2.15 is an illustration of a basic ultrafiltration set-up for waste minimization. However, ultrafiltration mechanism can also be implemented in some other set-up as well (Shahabadi and Reyhani, 2014).

2.7.4 Reverse osmosis

Reverse osmosis denotes the process that separates large molecules from waste using a semipermeable membrane against concentration gradient. That makes this process an energy driven process. Reverse osmosis is a largely used technique in multiple sectors. And in case of waste minimization, it can isolate pollutants using its basic principles. This process is often used for wastewater treatment where large and complex molecules should be isolated. Especially for colored effluent, reverse osmosis can isolate both color and salt compounds and thus, it can produce reusable clean water (Al-Bastaki, 2004). Even, this process reverse osmosis is found useful to isolate radioactive waste at an industrial level (Chmielewski et al., 2001). As a co-benefit of the process, several resources are often isolated using reverse osmosis. For instance, introducing reverse osmosis treatment in the waste streams of pulp industries could recover significant share of useful chemicals and reusable water. This minimized the resource buying cost for such industries significantly (Dubé and MacLatchy, 2000). Even tanning industries often adopt reverse osmosis as a tertiary treatment technology to minimize waste disposal by the industry (Bódalo et al., 2005). However, despite its usefulness in multiple sectors of waste minimization, as it is a pressure driven process, its energy efficiency is not often as much as expected.

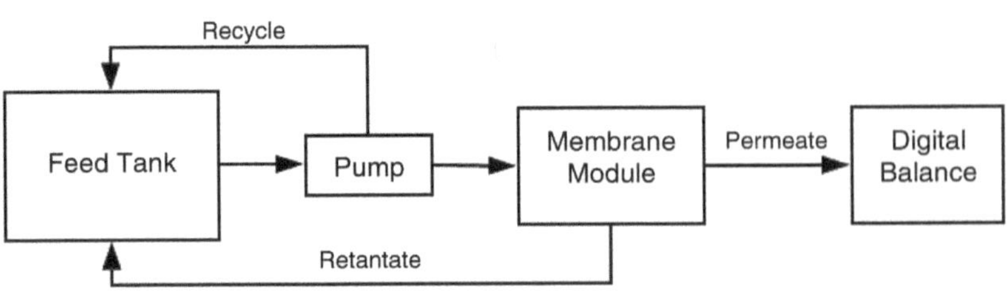

Figure 2.15 ***Schematic diagram showing common steps of ultrafiltration process.***

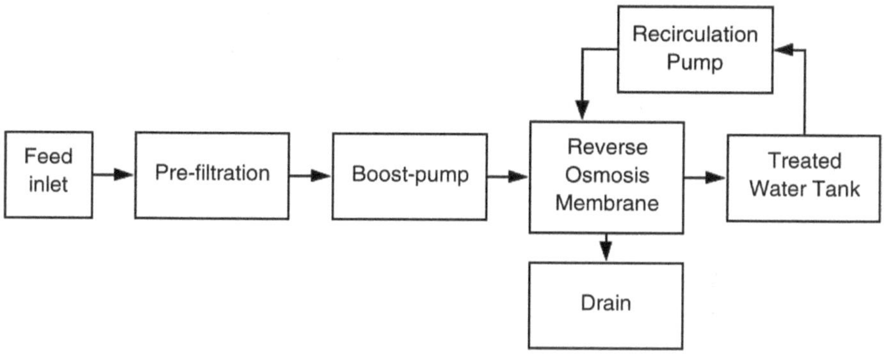

Figure 2.16 ***Wastewater treatment plant setup for reverse osmosis.***

In Fig. 2.16, a primary reverse osmosis set-up has been illustrated for waste minimization (Agarwal et al., 2016). As per the users' demand, extra facilities can be incorporated under this set-up.

2.7.5 Ocean cleanup system

Nowadays, ocean cleanup is an urgent need to ensure the sustainability of ocean biodiversity. Ocean is being polluted by disposal of industrial pollutants, oil spilt, and storage of layers of solid waste at ocean beds. Most importantly, plastic pollution has become a matter of huge concern nowadays, as plastics are stored in the ocean bed for a long time and harm growth of flora and fauna (Sigler, 2014). Realizing the harmful consequences of such waste accumulation in the ocean, U-shaped nets are often being used to clean oceans. These nets can clean large share of solid waste accumulated in ocean. Additionally, useful microorganisms are spread in the oceans to control ocean acidification and oil spill, as these microorganisms can neutralize ocean pH and can degrade the oil components accumulated in ocean surface (Redmond and Valentine, 2012) (Fig. 2.17).

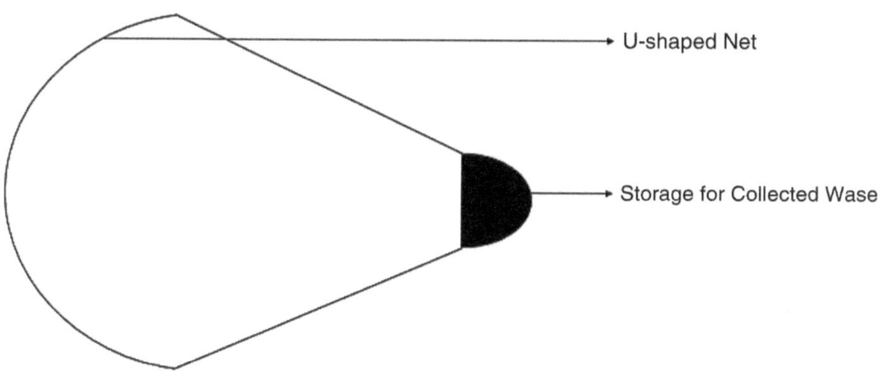

Figure 2.17 ***U-shaped nets to clean solid waste from ocean.***

2.8 Energy source modification

Modifying energy sources is a source level waste minimization strategy that reduces significant share of air, water, and soil pollution at pregeneration level. In this regard, limiting the use of fossil fuel is an urgent need. To do so, either renewable energy production should be expanded or fuels having less emission potential should be encouraged to use. Considering the wide range of fossil fuel usage, both of these tasks are quite challenging. But to minimize waste source, a massive modification in energy source can be a quite promising field to be considered.

2.8.1 Renewable energy generation

Renewable energy generation can be regarded as a source-level waste minimization strategy as it minimizes pollutant emission before its generation. Considering limited availability of fossil fuels and huge carbon emission through conventional energy generation system, more focus on renewable energy should be ensure to meet increasing energy demand using available resource. Besides, as renewable energy is ecofriendly in multiple ways, it is a more sustainable type of energy comparing with conventional energy generation processes. Moreover, even if renewable energy cannot minimize generation of all different types of waste materials, it can reduce the carbon footprints of different waste materials (Perry et al., 2008). In this regard, stakeholders from different sectors should be encouraged to utilize renewable energy; especially the industries should be instructed to have their own renewable energy plant through policy guideline (Wolsink, 2010). Thus, through policy regulation and industrial initiatives, renewable energy can be introduced as a waste source reduction strategy.

In this regard, solar energy, being the mostly available renewable resource, can be converted into energy using solar cells. Generation of solar energy emits almost no pollutant. In addition to it, if the solar energy producers produce extra energy, they can share their extra energy with others and thus, they can make profit as well. As a result, solar energy can be regarded as a sustainable waste minimization initiative. However, a basic pathway showing solar energy to usable solar fuel production (i.e., electricity) has been illustrated in Fig. 2.18 (Wang, 2020).

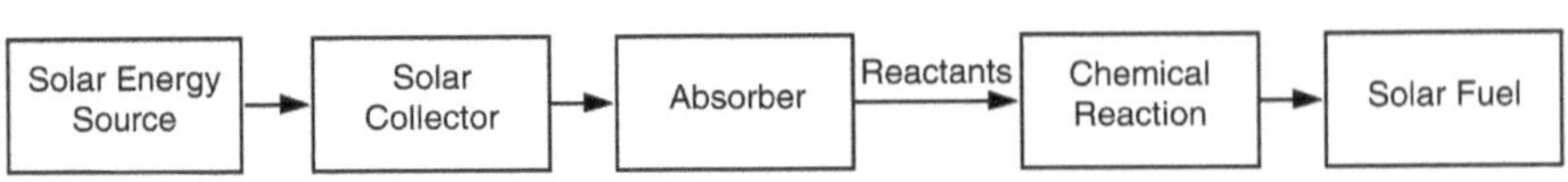

Figure 2.18 ***Solar thermochemical energy conversion.***

2.8.2 Introducing alternative energy sources

Finding alternative energy sources is an urgent need to minimize emission. Thus, fossil fuel-mediated emission will be regulated. To do so, either less emission resources or waste mediated energy production should be ensured. Among the resources having less emissions, radioactive materials, biogas, etc., are the notable ones which do not affect environment that negatively. Besides, several researches are going on currently to produce energy from waste at large scale. This supports the waste to energy concept of waste minimization strategy (Kothari et al., 2010). For example, green energy is often produced using waste biomass of municipal solid waste and palm oil (Moya et al., 2017; Yusoff, 2006). Besides, biogas is being introduced as an alternative energy source to meet increasing energy demand sustainably (Deublein and Steinhauser, 2011). Though these alternative energy sources are not 100% eco-friendly, but these have immense potentials to reduce pollutant emission and to minimize amount of produced waste. It is easier to implement utilization of such alternative resources at an industrial level, as industries often require to build their own strategies of energy generation and utilization (Kiran et al., 2014). This is why with the initiatives of the government, the industries, as the largest waste producers, should come forward to utilize more sustainable alternative energy sources. In addition to the waste to energy concept, overall waste management and energy recovery is required by implementing different waste reduction mechanisms. In doing so, long-term and large-scale initiatives of respective stakeholders are required.

Fig. 2.19 explores waste to energy mechanism incorporating energy recovery schemes. Understanding and implementing such closed loop mechanisms at different context simplifies full scheme of waste-energy relationship and increases yield/efficiency (Kariuki, 2016; Baran et al., 2016).

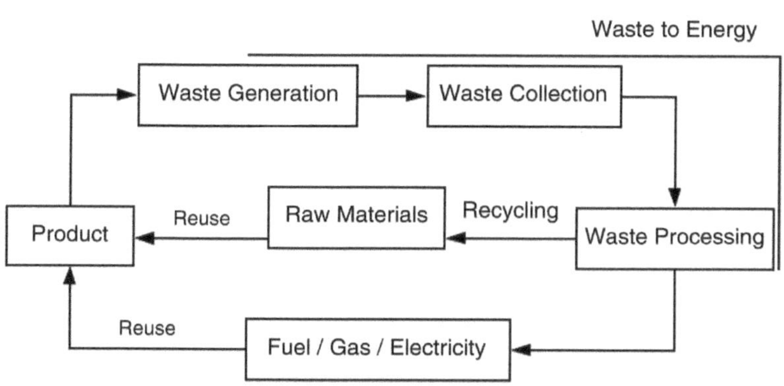

Figure 2.19 ***Illustration showing closed loop waste to energy mechanism.***

2.9 Information technology-based waste minimization

With the increasing use of smartphones and internet facility, information technology can now be considered as a constant companion of our everyday life. The diversity of the apps released for different purposes made people's task quite easier due to their properties of artificial intelligence and user-friendliness. Likewise, in waste minimization sector, utilization of different apps have impacted positively. Though most of the relevant apps used in this area, have a very narrow and selective focus, their performances benefitted individuals to regulate their waste footprint (Mavropoulos et al., 2015). Considering the diversity of the waste minimization issue, different apps have been launched to make it easier for the individuals and government in maintaining the issue more readily.

2.9.1 RFID tracking

RFID (radio frequency identification) tracking system can be regarded as a very useful waste minimization and waste management related soft tool. RFID tracking system helps to manage assets by transmitting electromagnetic information from assigned RFID tags of the waste storage to the corresponding readers. In case of waste minimization, RFID tracking system can be regarded as a less time consuming way for waste collection root optimization, waste transportation, waste minimization monitoring, and most importantly to smoothly manage the overall waste situation of an area (Arebey et al., 2009; Tang et al., 2008; Ustundag and Cevikcan, 2008). In many areas, using RFID tracking system, operation cost is optimized based on the amount of produced waste. It encourages individuals to minimize and recycle waste under a technology driven system (Ustundag and Cevikcan, 2008). Besides, using RFID tracking system, recycling merchants are notified to collect recyclable waste that ensures maximum recycling opportunities under a centralized and organized supply chain system of a municipality (Tang et al., 2008). This can also be applicable in large scale industrial setting to monitor, manage, and reduce industrial waste (Nunes et al., 2006). However, application of this tracking system is not very common in low income countries due to technological barrier and limited priority in waste minimization sector.

Fig. 2.20 shows RFID tracking system for waste collection process (Abdoli, 2009). In this process, the invoice is calculated as per amount of produced waste and geographic location of the collection spot. This ensures a systematic regulation of the whole loop.

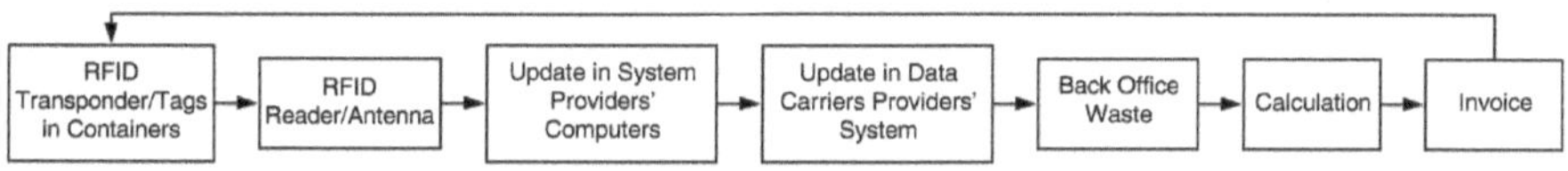

Figure 2.20 ***Workflow of RFID tracking system enabled for waste management.***

2.9.2 Apps and software

There are several apps and software that can track waste production and waste minimization initiatives by an individual. Most of such apps are designed to minimize food waste. In such regards, the apps like Sitebuddy and too good to go, etc. are often used to track individual's waste generation. For instance, the app, too good to go, claims itself as a "food waste warrior" who tracks the left-over food items of restaurants, bakeries, and supermarkets and distributes these food among the interested consumers at a cheaper price which is an integration of shared economy concept at food industry. This app has produced a win-win scheme for the consumers-retailers and most importantly for the environment. Even both the consumers and the retailers are quite satisfied for their economic gain and active contribution toward environmental benefit. Another notable app contributing actively in waste source reduction, Sitebuddy, is the online ordering platform of Reconomy, United Kingdom. This app is focused on construction waste minimization offering a skip hire facility to their customers. Personals and procurement teams working on house building and commercial site construction and management may send request to the authorizing company for skip hire service using the user friendly simple features of the app within few seconds. This app has successfully served the customers at the moment of any crucial need in terms of their construction waste management. Besides, the users found this app very useful in managing multiple sites and projects in an organized and timely manner. Thus, various companies are bringing about such innovative online platforms for sustainable waste source reduction considering the consumers' needs.

2.10 Automatic waste collection

One of the biggest obstacle, which is needed to be overcome to ensure smooth minimization of waste is to collect the waste efficiently. Especially in case of the municipalities, it is tough to ensure harmless collection and transportation of waste from households and industries to the treatment plant. To solve this problem, automatic waste collection is being introduced in several areas that are monitored under a centralized system of waste management and treatment.

Pneumatic tube system is an automatic waste collection system that uses vacuum to collect the waste materials from the connect tube system (Hidalgo et al., 2018). This is quite similar to a gripper system of waste collection that collects waste in an automated way through tube system (Curotto, 2014). These technologies make the task easier to be smoothly maintained and prohibit exposure of waste in open space. The major challenge of this process is to afford the huge cost of the technology installation, operation, and maintenance.

In this regard, the example of logiwaste can be quite relevant. Logiwaste offers automated waste collection system using pipeline system where waste materials are transported from a disposal inlet to a centralized terminal. In this process, vacuum in pipeline system helps to transport waste materials efficiently. Then the authority of centralized terminal adopts strategies to treat and minimize waste. So far, Logiwaste has successfully implement such projects in both large and small scale using their automated waste collection system at an affordable cost and efficient mechanism. Their automated Waste Collection System includes different steps such as (1) Waste Disposal Inlet; (2) Waste Staging; (3) Tube System for Waste Collection; (4) Waste Collection in the Container at Terminal; (5) Waste Transportation System at the Terminal (LogiWaste, 2020).

2.11 Green chemistry: as a waste minimization strategy

Green chemistry is an emerging field that designs chemical products and chemical processes with an aim to eliminate the generation and usage of harmful hazardous substances. The concept green chemistry works under 12 major principles among which the first principle is "Waste Prevention." Likewise, waste minimization strategies also support this principle. Even the other 11 principles of green chemistry also suggests implementation of waste minimization principles from a chemical and analytical point of view. It is an indicator that green chemistry and waste minimization strategies can be aligned together to achieve a better outcome. Both green chemistry and waste minimization strategies discourage utilization of hazardous waste in industrial process designing as an environmental safety measurement (Mooney, 2004). Besides, green chemistry measurements help industrial process designing to adopt cleaner strategies and safer chemicals that do not produce much waste at the end of production and manufacturing (Pandian et al., 2009). Thus, green chemistry is often regarded as a useful waste minimization strategy. Here, Table 2.1 discusses and explores the principles of green chemistry in association with waste minimization and source reduction approaches.

Table 2.1 Connecting 12 principles of green chemistry with waste minimization and source reduction strategies (de Marco et al., 2019; Anastas and Warner, 1998; Compound Interest, 2015).

Principles of green chemistry	Waste minimization and source reduction strategies
Principle 1: Prevention	Waste prevention prioritizing pregeneration treatment through multiple approaches, that is, process modification, recycling, treatment, etc.
Principle 2: Atom economy	Waste reduction at an atomic/molecular level or evaluating reaction efficiency ensuring waste reduction.
Principle 3: Less hazardous chemical synthesis	Increased efficiency of chemical reactions and synthetic routes help to minimize waste.
Principle 4: Safer chemicals design	Molecular modifications of chemicals minimize toxicity of waste and its harmful consequences.
Principle 5: Use of safer solvents & auxiliaries	Solvents being a major example of waste, choosing the safest solvent reduce amount, and negative impacts of generated waste.
Principle 6: Energy efficiency	Increased energy efficiency reduces energy mediated waste production. Solar energy consumption, waste to energy mechanisms both reduce waste and increase energy efficiency.
Principle 7: Use of renewable raw materials	Renewable materials are found to be significantly less-waste producing during and after the process.
Principle 8: Reduction of derivatives	Avoiding derivative generation during reactions ensures less number of reaction steps, leading less waste generation.
Principle 9: Catalysis	Catalysis increases energy and resource efficiency assuring a significant reduction of waste generation.
Principle 10: Product design for degradation	Such product designing aims degradation-oriented end-waste for which such less-persistent waste also ensure source-level waste reduction within shortest time of waste generation.
Principle 11: Real-time analysis for pollution prevention	Real-time pollution prevention reduces waste simultaneous to the chemical reactions that leads to zero waste production during industrial process.
Principle 12: Accident prevention	Such modifications under green chemistry principle, prohibit waste-borne hazards/accidents underlining potential risks beforehand which is also a source-level waste minimization strategy.

As mentioned in Table 2.1, all these green chemistry principles somehow encourage waste minimization using chemical techniques instead of encouraging treatment of hazardous materials. In addition to principle based waste minimization scenario presented in Table 2.1, there are broad strategies that suggest green chemistry-mediated waste minimization. First, approaches of green chemistry can be used to retrieve useful resources from waste materials for which approaches of green chemistry can be considered for waste to wealth generation which is a crucial waste minimization strategy. For example, energy generation from biomass derived from waste production requires intrinsic chemical analysis for which approaches of green chemistry can be useful (Clark et al., 2013). In addition to it, green chemistry also supports energy savings by encouraging catalysis for chemical reaction. And energy saving is necessary for pollution control at a mass level. Thus, catalysis, being a basic principle of green chemistry, may contribute in pollution control and waste minimization (Anastas et al., 2001). Moreover, as ensuring sustainability is a major focus of green chemistry, to ensure long term sustainability, waste source must be minimized adopting several approaches of green chemistry. Most importantly, all the chemical processes required for minimization of waste generation requires implications of green chemistry principles (Clark and Macquarrie, 2008). Thus, adoption of several strategies supported by green chemistry helps to design waste minimization initiatives successfully. In other word, green chemistry will thus also help to promote source-level waste minimization.

In addition to green chemistry principles, principles of green engineering can also be useful to design sustainable waste minimization technologies, as approaches of green engineering contributes in beneficial synergies (Anastas and Zimmerman, 2006). Green engineering being yet an emerging field of sustainable process and is designing acts efficiently like the green chemistry approaches (Table 2.2).

Despite having several overlapping objectives and principles in between green chemistry and green engineering, green engineering is comparatively a broader cluster of criteria to be followed for sustainable process designing. Applications of green engineering in waste source reduction mechanism minimize detrimental environmental consequences of all the processes involved with manufacture and operations (Davis, 2005). This is how green engineering approaches and technologies modify industrial processes sustainably.

Table 2.2 12 Principles of green engineering connecting with waste minimization and source reduction (Anastas and Zimmerman, 2006).

Principle	Descriptions
Principle 1:	All materials and energy sources should be designed such that they are non-hazardous.
Principle 2:	Preventing waste should be prioritized over treatment or clean up.
Principle 3:	Inclusive design should be introduced incorporating separation and purification operations
Principle 4:	Maximum efficiency should be assured by the system.
Principle 5:	System components should be output pulled instead of being input pushed through proper use of resources.
Principle 6:	Complexity associated with recycle, reuse, and beneficial deposition should be regarded as an investment.
Principle 7:	Design goal should include durability potentials in lieu of immortality.
Principle 8:	Unnecessary capacity of system is a design flaw.
Principle 9:	Material diversity should be minimized.
Principle 10:	Interconnectivity of processes and systems should be ensured using limited energy and materials.
Principle 11:	"After-life" performance in commercial level needs to be analyzed.
Principle 12:	Renewable and readily inputs should be incorporated in technology designing.

Case Study 1: "Waste to Wealth"—Utilization of Coconut Shell for Plantation

Waste minimization approach can be as simple as reusing a waste material without any sort of modification to as complex as processes like electrolysis. Whatever the process is, minimizing waste is the major goal of all such concepts. In this regard, example of using tender coconut shells for plantation can be a real-life example of waste minimization at a household to local level. Coconut shells are anti-bacterial and anti-viral components, which can protect plants from different bacterial and viral infections and thus it can ensure smooth growth of the plants. Besides, it has its aesthetic components as well that increases beauty of gardens and households. This is why planting trees in coconut shell is an emerging idea in India. Besides when such plants are transferred to soil along with the shells, soil fertility increases through nutrient release after decomposition of the coconut shell. This simple strategy can be adopted in the other parts of the world to minimize improper disposal of the shell.

Case Study 2: Food Waste Recycling—South Korea's Effort to Become a Zero Waste Society

Food waste, being a daily generated waste, needs to be minimized drastically to be a zero waste society. In this regard, the effort of South Korea is a very encouraging story for societies that are struggling to become zero waste societies. Gradually initiatives adopted by the government and the individuals to be a zero waste society, helped the country to recycle more than 95% of their food wastes today. Even in 1995, only 2% of the food waste was recycled in South Korea. So, it is quite surprising to detect what made them to be so successful in recycling food waste efficiently.

However, there are a series of initiatives which helped South Korea to be such a successful food waste recycler. By imposing acts and introducing policies, South Korea could ensure successful recycling. The government introduced Waste Control Act, Refuse Cleaning Act, Environment Conservation Act, and Recycling Promotion Act at the end of last century. In addition to it, they introduced Zero Waste policies, Master Plans for Waste Management, and several Frameworks. All the industries are responsible to strictly maintain the series of acts, policies, plans, and frameworks. Smoothly regulated waste collection and treatment technology were also ensured. Additionally, the government and organizations were promoting smart bins, urban farms, and other practices that encouraged the country people to adopt and maintain these voluntarily. The education system also incorporated basic concepts and contents of waste minimization (Seoul Metropolitan Government, 2016).

Thus, the urban farms and even the rural farmers were motivated to compost their food waste instead of disposing it without any sort of treatment. Even the industries and waste treatment plants started recovering useful resources by treating food waste. As a result, a significant increase in terms of food waste recycling was noticed countrywide.

References

Abdoli, S. (2009). RFID application in municipal soli waste management system. *International Journal of Environmental Research, 3*(3), 447–454.

Achinas, S., & Kapetanios, E. (2013). Efficiency evaluation of RDF plasma gasification process. *Energy and Environment Research, 3*(1), 150.

Agarwal, M., Singh, K., Dohare, R. K., & Upadhyaya, S. (2016). Process control and optimization of wastewater treatment plants using simulation softwares: a review. *International Journal of Advanced Technology and Engineering Exploration, 3*(22), 145.

Al-Bastaki, N. (2004). Removal of methyl orange dye and Na_2SO_4 salt from synthetic waste water using reverse osmosis. *Chemical Engineering and Processing: Process Intensification, 43*(12), 1561–1567.

Anastas, P.T., & Warner, J. C. (1998). Principles of green chemistry. *Green chemistry: Theory and practice*. New York: Oxford University Press. 29-56.

Anastas, P. T., & Zimmerman, J. B. (2006). The twelve principles of green engineering as a foundation for sustainability. In *Sustainability science, engineering* (Vol. 1). Elsevier. 11-32.

Anastas, P. T., Kirchhoff, M. M., & Williamson, T. C. (2001). Catalysis as a foundational pillar of green chemistry. *Applied Catalysis A: General, 221*(1–2), 3–13.

Andreola, N.M., Barbieri, L., Lancellotti, I., & Pozzi, P. (2005). Recycling industrial waste in brick manufacture. Part 1. *Materiales De Construcción, 55*(280), 5–16.

Appels, L., Baeyens, J., Degrève, J., & Dewil, R. (2008). Principles and potential of the anaerobic digestion of waste-activated sludge. *Progress in Energy and Combustion Science, 34*(6), 755–781.

Arebey, M., Hannan, M.A., Basri, H., & Abdullah, H. (2009). Solid waste monitoring and management using RFID, GIS and GSM. In *2009 IEEE student conference on research and development (SCOReD)*, (pp. 37–40). IEEE.

Armstrong, B. A., & Reinhardt, P. A. (2010). Managing laboratory biomedical waste using a large on-site autoclave–shredder. *Journal of Chemical Health & Safety, 17*(6), 33–39.

Assamoi, B., & Lawryshyn, Y. (2012). The environmental comparison of landfilling vs. incineration of MSW accounting for waste diversion. *Waste Management, 32*(5), 1019–1030.

Aziz, H. A., Adlan, M. N., Zahari, M. S. M., & Alias, S. (2004). Removal of ammoniacal nitrogen ($N-NH_3$) from municipal solid waste leachate by using activated carbon and limestone. *Waste Management & Research, 22*(5), 371–375.

Baran, B., Mamis, M. S., & Alagoz, B. B. (2016). Utilization of energy from waste potential in Turkey as distributed secondary renewable energy source. *Renewable Energy, 90*, 493–500.

Bartoli, M., Rosi, L., Frediani, M., Undri, A., & Frediani, P. (2015). Depolymerization of polystyrene at reduced pressure through a microwave assisted pyrolysis. *Journal of Analytical and Applied Pyrolysis, 113*, 281–287.

Basso, D., Patuzzi, F., Castello, D., Baratieri, M., Rada, E. C., Weiss-Hortala, E., et al. (2016). Agro-industrial waste to solid biofuel through hydrothermal carbonization. *Waste Management, 47*, 114–121.

Berge, N. D., Ro, K. S., Mao, J., Flora, J. R.V., Chappell, M. A., & Bae, S. (2011). Hydrothermal carbonization of municipal waste streams. *Environmental Science & Technology, 45*(13), 5696–5703.

Bhatluri, K. K., Manna, M. S., Ghoshal, A. K., & Saha, P. (2015). Supported liquid membrane based removal of lead (II) and cadmium (II) from mixed feed: conversion to solid waste by precipitation. *Journal of Hazardous Materials, 299*, 504–512.

Bielska, M., & Szymanowski, J. (2006). Removal of methylene blue from waste water using micellar enhanced ultrafiltration. *Water Research, 40*(5), 1027–1033.

Biller, P., & Ross, A. B. (2016). Production of biofuels via hydrothermal conversion. *Handbook of biofuels production*. Woodhead Publishing. 509-547.

Bódalo, A., Gómez, J. L., Gómez, E., Hidalgo, A. M., & Alemán, A. (2005). Viability study of different reverse osmosis membranes for application in the tertiary treatment of wastes from the tanning industry. *Desalination, 180*(1–3), 277–284.

Brissaud, F. (2008). Criteria for water recycling and reuse in the Mediterranean countries. *Desalination, 218*(1–3), 24–33.

Cao, L., Zhang, C., Chen, H., Tsang, D. C., Luo, G., Zhang, S., et al. (2017). Hydrothermal liquefaction of agricultural and forestry wastes: state-of-the-art review and future prospects. *Bioresource Technology, 245*, 1184–1193.

Chen, D., Yin, L., Wang, H., & He, P. (2014). Pyrolysis technologies for municipal solid waste: a review. *Waste Management, 34*(12), 2466–2486.

Chinnaraj, S., & Rao, G.V. (2006). Implementation of an UASB anaerobic digester at bagasse-based pulp and paper industry. *Biomass and Bioenergy, 30*(3), 273–277.

Chmielewski, A. G., Harasimowicz, M., Tyminski, B., & Zakrzewska-Trznadel, G. (2001). Concentration of low-and medium-level radioactive wastes with three-stage reverse osmosis pilot plant. *Separation Science and Technology*, *36*(5–6), 1117–1127.

Clark, J. H., & Macquarrie, D. J. (Eds.). (2008). *Handbook of green chemistry and technology*. John Wiley & Sons.

Clark, J. H., Pfaltzgraff, L. A., Budarin, V. L., Hunt, A. J., Gronnow, M., Matharu, A. S., et al. (2013). From waste to wealth using green chemistry. *Pure and Applied Chemistry*, *85*(8), 1625–1631.

Compound Interest (2015). *Environmental chemistry: the twelve principles of green chemistry: what it is & why it matters*. https://www.compoundchem.com/2015/09/24/green-chemistry/.

Curotto, J. M. (2014). *U.S. Patent No. 8,753,062*. Washington, DC: U.S. Patent and Trademark Office.

Davis, A. P. (2005). Green engineering principles promote low-impact development. *Environmental Science & Technology*, *39*(16), 338A–344B.

De Gioannis, G., Muntoni, A., Cappai, G., & Milia, S. (2009). Landfill gas generation after mechanical biological treatment of municipal solid waste. Estimation of gas generation rate constants. *Waste Management*, *29*(3), 1026–1034.

de Marco, B. A., Rechelo, B. S., Tótoli, E. G., Kogawa, A. C., & Salgado, H. R. N. (2019). Evolution of green chemistry and its multidimensional impacts: A review. *Saudi Pharmaceutical Journal*, *27*(1), 1–8.

Deleris, S., Geaugey, V., Camacho, P., Debellefontaine, H., & Paul, E. (2002). Minimization of sludge production in biological processes: an alternative solution for the problem of sludge disposal. *Water Science and Technology*, *46*(10), 63–70.

Deublein, D., & Steinhauser, A. (2011). *Biogas from waste and renewable resources: an introduction*. John Wiley & Sons.

Dimitriadis, A., & Bezergianni, S. (2017). Hydrothermal liquefaction of various biomass and waste feedstocks for biocrude production: A state of the art review. *Renewable and Sustainable Energy Reviews*, *68*, 113–125.

Dubé, M. G., & MacLatchy, D. L. (2000). Endocrine responses of Fundulus heteroclitus to effluent from a bleached-kraft pulp mill before and after installation of reverse osmosis treatment of a waste stream. *Environmental Toxicology and Chemistry: An International Journal*, *19*(11), 2788–2796.

Emek, E., & Kara, B. Y. (2007). Hazardous waste management problem: The case for incineration. *Computers & Operations Research*, *34*(5), 1424–1441.

Flandinet, L., Tedjar, F., Ghetta, V., & Fouletier, J. (2012). Metals recovering from waste printed circuit boards (WPCBs) using molten salts. *Journal of Hazardous Materials*, *213*, 485–490.

Galeno, G., Minutillo, M., & Perna, A. (2011). From waste to electricity through integrated plasma gasification/fuel cell (IPGFC) system. *International Journal of Hydrogen Energy*, *36*(2), 1692–1701.

Guedes, R. E., Luna, A. S., & Torres, A. R. (2018). Operating parameters for bio-oil production in biomass pyrolysis: A review. *Journal of Analytical and Applied Pyrolysis*, *129*, 134–149.

Halim, I., & Srinivasan, R. (2002). Systematic waste minimization in chemical processes. 1. Methodology. *Industrial & Engineering Chemistry Research*, *41*(2), 196–207.

Hamilton, D. W. (2004). Methods for determining anaerobic/facultative lagoon operation life using a complex sludge accumulation model. In *2004 ASAE Annual Meeting* (p. 1). American Society of Agricultural and Biological Engineers.

Hasan, M. M., Hossain, M. J., Chowdhury, S., Halim, M. I., Marjia, U., & Rashid, H. A. (2017). A novel proposition of facilities required for sterile pharmaceutical preparation. *The Pharma Innovation Journal*, *6*(10), 125–146.

Hidalgo, D., Martín-Marroquín, J. M., Corona, F., & Juaristi, J. L. (2018). Sustainable vacuum waste collection systems in areas of difficult access. *Tunnelling and Underground Space Technology*, *81*, 221–227.

Hossain, M., Balakrishnan, V., Rahman, N. N. N. A., Sarker, M., Islam, Z., & Kadir, M. O. A. (2012). Treatment of clinical solid waste using a steam autoclave as a possible alternative technology to incineration. *International Journal of Environmental Research and Public Health*, *9*(3), 855–867.

Hsu, P. C., Foster, K. G., Ford, T. D., Wallman, P. H., Watkins, B. E., Pruneda, C. O., et al. (2000). Treatment of solid wastes with molten salt oxidation. *Waste Management*, *20*(5–6), 363–368.

Hu, B., Wang, K., Wu, L., Yu, S. H., Antonietti, M., & Titirici, M. M. (2010). Engineering carbon materials from the hydrothermal carbonization process of biomass. *Advanced Materials*, *22*(7), 813–828.

Huang, X., Wang, W., Liu, Y., Wang, H., Zhang, Z., Fan, W., et al. (2015). Treatment of oily waste water by PVP grafted PVDF ultrafiltration membranes. *Chemical Engineering Journal*, *273*, 421–429.

Huang, Y. C., Batchelor, B., & Koseoglu, S. S. (1994). Crossflow surfactant-based ultrafiltration of heavy metals from waste streams. *Separation Science and Technology*, *29*(15), 1979–1998.

Ioana, I. (2010). Clean technology from waste management. In *Advances in Waste Management, fourth WSEAS International Conference on Waste Management, Water Pollution, Air Pollution, Indoor climate (WWAI'10)*, (pp. 155–171).

Jäckel, U., Thummes, K., & Kämpfer, P. (2005). Thermophilic methane production and oxidation in compost. *FEMS Microbiology Ecology*, *52*(2), 175–184.

Kadier, A., Simayi, Y., Abdeshahian, P., Azman, N. F., Chandrasekhar, K., & Kalil, M. S. (2016). A comprehensive review of microbial electrolysis cells (MEC) reactor designs and configurations for sustainable hydrogen gas production. *Alexandria Engineering Journal*, *55*(1), 427–443.

Karayildirim, T., Yanik, J., Yuksel, M., & Bockhorn, H. (2006). Characterisation of products from pyrolysis of waste sludges. *Fuel*, *85*(10–11), 1498–1508.

Kariuki, D. (2016). Recovering Energy from Waste Can Power Africa, *CleanLeap.com*.

Khalid, A., Arshad, M., Anjum, M., Mahmood, T., & Dawson, L. (2011). The anaerobic digestion of solid organic waste. *Waste Management*, *31*(8), 1737–1744.

Kiran, E. U., Trzcinski, A. P., Ng, W. J., & Liu, Y. (2014). Bioconversion of food waste to energy: a review. *Fuel*, *134*, 389–399.

Kothari, R., Tyagi, V. V., & Pathak, A. (2010). Waste-to-energy: A way from renewable energy sources to sustainable development. *Renewable and Sustainable Energy Reviews*, *14*(9), 3164–3170.

Kumar, A., Dewulf, J., & Van Langenhove, H. (2008). Membrane-based biological waste gas treatment. *Chemical Engineering Journal*, *136*(2–3), 82–91.

Kumar, T. P., Rahul, M. A., & Chandrajit, B. (2011). Biofiltration of volatile organic compounds (VOCs): an overview. *Research Journal of Chemical Sciences*, *2231*, 606X.

Kuo, J. H., Tseng, H. H., Rao, P. S., & Wey, M. Y. (2008). The prospect and development of incinerators for municipal solid waste treatment and characteristics of their pollutants in Taiwan. *Applied Thermal Engineering*, *28*(17–18), 2305–2314.

Kwon, H. B., Lee, C. W., Jun, B. S., Weon, S. Y., & Koopman, B. (2004). Recycling waste oyster shells for eutrophication control. *Resources, Conservation and Recycling*, *41*(1), 75–82.

Li, Y. F., Yao, Y., & Wang, L. (2009). Recycling of industrial waste and performance of steel slag green concrete. *Journal of Central South University of Technology*, *16*(5), 768.

Liang, C., Das, K. C., & McClendon, R. W. (2003). The influence of temperature and moisture contents regimes on the aerobic microbial activity of a biosolids composting blend. *Bioresource Technology*, *86*(2), 131–137.

LogiWaste (2020). How waste management works with the Logiwaste automated vacuum waste collection system. *LogiWaste.com.*

Lu, L., & Ren, Z. J. (2016). Microbial electrolysis cells for waste biorefinery: A state of the art review. *Bioresource Technology, 215*, 254–264.

Lu, X., Jordan, B., & Berge, N. D. (2012). Thermal conversion of municipal solid waste via hydrothermal carbonization: Comparison of carbonization products to products from current waste management techniques. *Waste Management, 32*(7), 1353–1365.

Macedonio, F., Brunetti, A., Barbieri, G., & Drioli, E. (2012). Membrane condenser as a new technology for water recovery from humidified "waste" gaseous streams. *Industrial & Engineering Chemistry Research, 52*(3), 1160–1167.

Maddi, B., Panisko, E., Wietsma, T., Lemmon, T., Swita, M., Albrecht, K., et al. (2017). Quantitative characterization of aqueous byproducts from hydrothermal liquefaction of municipal wastes, food industry wastes, and biomass grown on waste. *ACS Sustainable Chemistry & Engineering, 5*(3), 2205–2214.

Marculescu, C. (2011). Thermal-chemical treatment of solid waste mixtures. *Energy Procedia, 6*, 558–564.

Martínez, J. D., Puy, N., Murillo, R., García, T., Navarro, M.V., & Mastral, A. M. (2013). Waste tyre pyrolysis–A review. *Renewable and Sustainable Energy Reviews, 23*, 179–213.

Mavropoulos, A., Tsakona, M., & Anthouli, A. (2015). Urban waste management and the mobile challenge. *Waste Management & Research, 33*(4), 381–387.

McNEVIN, D., & Barford, J. (2000). Biofiltration as an odour abatement strategy. *Biochemical Engineering Journal, 5*(3), 231–242.

Meiqiang, L.Y.Z.Y.L. (2003). The treatment of dyeing wastewater by the electrocoagulation-biological activated carbon filtering reactor process [J]. *Techniques and Equipment for Environmental Pollution Control, 2.*

Melikoglu, M., Lin, C. S. K., & Webb, C. (2013). Kinetic studies on the multi-enzyme solution produced via solid state fermentation of waste bread by Aspergillus awamori. *Biochemical Engineering Journal, 80*, 76–82.

Misra, R. V., Roy, R. N., & Hiraoka, H. (2003). *On-farm composting methods.* Rome, Italy: UN-FAO.

Mooney, D. (2004). Effectively minimizing hazardous waste in academia: The green chemistry approach. *Chemical Health & Safety, 11*(3), 24–28.

Moya, D., Aldás, C., López, G., & Kaparaju, P. (2017). Municipal solid waste as a valuable renewable energy resource: a worldwide opportunity of energy recovery by using waste-to-energy technologies. *Energy Procedia, 134*, 286–295.

Narayan, A. V., & Agrawal, P. (2012). Enzyme based processes for removal of phenol from waste water: current status and future challenges. *Journal of Environmental Research and Development*, 7(2), 724–728.

Nicolae, S. A., Au, H., Modugno, P., Luo, H., Szego, A. E., Qiao, M., et al. (2020). Recent advances in hydrothermal carbonisation: from tailored carbon materials and biochemicals to applications and bioenergy. *Green Chemistry, 22*(15), 4747–4800. doi: 0.1039/D0GC00998A.

Nielsen, M.C., DePuy, R.A., Kumar, A., Lyons, J.P.F., Lissianski, V., Shi, R., et al. (2015). *U.S. Patent No. 9,074,152.* Washington, DC: U.S. Patent and Trademark Office.

Nunes, K. R., Schnatmeyer, M., Thoben, K. D., & Valle, R. A. (2006). Using RFID for waste minimization in the automotive industry. *IFAC Proceedings Volumes, 39*(3), 221–226.

Nyashina, G. S., Vershinina, K.Y., Shlegel, N. E., & Strizhak, P.A. (2019). Effective incineration of fuel-waste slurries from several related industries. *Environmental Research, 176*, 108559.

Olivieri, A. W., Seto, E., Cooper, R. C., Cahn, M. D., Colford, J., Crook, J., et al. (2014). Risk-based review of California's water-recycling criteria for agricultural irrigation. *Journal of Environmental Engineering, 140*(6), 04014015.

Pandian, S. R. K., Deepak, V., Kalishwaralal, K., Muniyandi, J., Rameshkumar, N., & Gurunathan, S. (2009). Synthesis of PHB nanoparticles from optimized medium utilizing dairy industrial waste using Brevibacterium casei SRKP2: a green chemistry approach. *Colloids and Surfaces B: Biointerfaces, 74*(1), 266–273.

Parales, R. E., Bruce, N. C., Schmid, A., & Wackett, L. P. (2002). Biodegradation, biotransformation, and biocatalysis (B3). *Applied and Environmental Microbiology, 68*(10), 4699–4709.

Paul, E., & Liu, Y. (Eds.). (2012). *Biological sludge minimization and biomaterials/bioenergy recovery technologies*. John Wiley & Sons.

Peigné, J., & Girardin, P. (2004). Environmental impacts of farm-scale composting practices. *Water, Air, and Soil Pollution, 153*(1–4), 45–68.

Pelino, M., Cantalini, C., & Rincon, J. M. (1997). Preparation and properties of glass-ceramic materials obtained by recycling goethite industrial waste. *Journal of Materials Science, 32*(17), 4655–4660.

Perry, S., Klemeš, J., & Bulatov, I. (2008). Integrating waste and renewable energy to reduce the carbon footprint of locally integrated energy sectors. *Energy, 33*(10), 1489–1497.

Ramke, H. G., Blöhse, D., Lehmann, H. J., & Fettig, J. (2009). Hydrothermal carbonization of organic waste. In Cossu, R., Diaz, LF, Stegman, R. (Eds.), *Twelfth international waste management and landfill symphosium*. Sardina: Pro., CISA pub.

Raza, W., Lee, J., Raza, N., Luo, Y., Kim, K. H., & Yang, J. (2019). Removal of phenolic compounds from industrial waste water based on membrane-based technologies. *Journal of Industrial and Engineering Chemistry, 71*, 1–18.

Redmond, M. C., & Valentine, D. L. (2012). Natural gas and temperature structured a microbial community response to the Deepwater Horizon oil spill. *Proceedings of the National Academy of Sciences, 109*(50), 20292–20297.

Richard, T. L., Hamelers, H. V. M., Veeken, A., & Silva, T. (2002). Moisture relationships in composting processes. *Compost Science & Utilization, 10*(4), 286–302.

Rodriguez, C., Hiligsmann, S., Ongena, M., Charlier, R., & Thonart, P. (2005). Development of an enzymatic assay for the determination of cellulose bioavailability in municipal solid waste. *Biodegradation, 16*(5), 415–422.

Sanlisoy, A., & Carpinlioglu, M. O. (2017). A review on plasma gasification for solid waste disposal. *International Journal of Hydrogen Energy, 42*(2), 1361–1365.

Santoleri, J. J., Reynolds, J., & Theodore, L. (2000). *Introduction to hazardous waste incineration*. John Wiley & Sons.

Scala, F. (2001). Simulation of mercury capture by activated carbon injection in incinerator flue gas. 2. Fabric filter removal. *Environmental Science & Technology, 35*(21), 4373–4378.

Seoul Metropolitan Government (2016). Minimizing food waste: zero food waste, *seoulsolution.kr.*

Shahabadi, S. M. S., & Reyhani, A. (2014). Optimization of operating conditions in ultrafiltration process for produced water treatment via the full factorial design methodology. *Separation and Purification Technology, 132*, 50–61.

Sigler, M. (2014). The effects of plastic pollution on aquatic wildlife: current situations and future solutions. *Water, Air, & Soil Pollution, 225*(11), 2184.

Silva, M. R., Coelho, M. A. Z., & Araújo, O. Q. F. (2002). Minimization of phenol and ammoniacal nitrogen in refinery wastewater employing biological treatment. *Revista de Engenharia Térmica, 1*(2), 33–37.

Suriani, A. B., Nor, R. M., & Rusop, M. (2010). Vertically aligned carbon nanotubes synthesized from waste cooking palm oil. *Journal of the Ceramic Society of Japan, 118*(1382), 963–968.

Tang, L. L., Lu, M.T., & Hung, W. C. (2008). RFID application in waste monitoring management of electronic products. *IJEBM*, *6*(3), 161–173.
Tiller, T., & Linscott, A. (2004). Evaluation of a steam autoclave for sterilizing medical waste at a university health center. *American Journal of Infection Control*, *32*(3), E9.
Tomar, P., & Suthar, S. (2011). Urban wastewater treatment using vermi-biofiltration system. *Desalination*, *282*, 95–103.
Tortora, F., Innocenzi, V., Prisciandaro, M., Vegliò, F., & Di Celso, G. M. (2016). Heavy metal removal from liquid wastes by using micellar-enhanced ultrafiltration. *Water, Air, & Soil Pollution*, *227*(7), 240.
Troschinetz, A. M., & Mihelcic, J. R. (2009). Sustainable recycling of municipal solid waste in developing countries. *Waste Management*, *29*(2), 915–923.
Ustundag, A. L. P., & Cevikcan, E. (2008). Vehicle route optimization for RFID integrated waste collection system. *International Journal of Information Technology & Decision Making*, 7(04), 611–625.
Van Zwieten, L., Kimber, S., Morris, S., Chan, K.Y., Downie, A., Rust, J., et al. (2010). Effects of biochar from slow pyrolysis of paper mill waste on agronomic performance and soil fertility. *Plant and Soil*, *327*(1–2), 235–246.
Vlyssides, A., Karlis, P., Loizidou, M., Zorpas, A., & Arapoglou, D. (2001). Treatment of leachate from a domestic solid waste sanitary landfill by an electrolysis system. *Environmental Technology*, *22*(12), 1467–1476.
Wang, H. (2020). Solar thermochemical fuel generation. In *Wind solar hybrid renewable energy system*. IntechOpen.
Wei, Y., Van Houten, R. T., Borger, A. R., Eikelboom, D. H., & Fan, Y. (2003). Minimization of excess sludge production for biological wastewater treatment. *Water Research*, *37*(18), 4453–4467.
Wibowo, S., & Deng, H. (2015). Multi-criteria group decision making for evaluating the performance of e-waste recycling programs under uncertainty. *Waste Management*, *40*, 127–135.
Wilson, D. C., Velis, C., & Cheeseman, C. (2006). Role of informal sector recycling in waste management in developing countries. *Habitat International*, *30*(4), 797–808.
Wolsink, M. (2010). Contested environmental policy infrastructure: Socio-political acceptance of renewable energy, water, and waste facilities. *Environmental Impact Assessment Review*, *30*(5), 302–311.
Yao, Z., Li, J., & Zhao, X. (2011). Molten salt oxidation: A versatile and promising technology for the destruction of organic-containing wastes. *Chemosphere*, *84*(9), 1167–1174.
Yusoff, S. (2006). Renewable energy from palm oil–innovation on effective utilization of waste. *Journal of Cleaner Production*, *14*(1), 87–93.
Zhang, R., El-Mashad, H. M., Hartman, K., Wang, F., Liu, G., Choate, C., et al. (2007). Characterization of food waste as feedstock for anaerobic digestion. *Bioresource Technology*, *98*(4), 929–935.
Zhang, Y., & Angelidaki, I. (2014). Microbial electrolysis cells turning to be versatile technology: recent advances and future challenges. *Water Research*, *56*, 11–25.
Zhao, L., Wang, Y., Yang, J., Xing, M., Li, X., Yi, D., et al. (2010). Earthworm–microorganism interactions: a strategy to stabilize domestic wastewater sludge. *Water Research*, *44*(8), 2572–2582.
Zhen, G.Y., Lu, X. Q., Li, Y.Y., & Zhao, Y. C. (2013). Innovative combination of electrolysis and Fe (II)-activated persulfate oxidation for improving the dewaterability of waste activated sludge. *Bioresource Technology*, *136*, 654–663.
Zsigraiová, Z., Tavares, G., Semiao, V., & de Graça Carvalho, M. (2009). Integrated waste-to-energy conversion and waste transportation within island communities. *Energy*, *34*(5), 623–635.

Further reading

Sheldon, R. A. (2017). The E factor 25 years on: the rise of green chemistry and sustainability. *Green Chemistry, 19*(1), 18–43.

Sitkey, V., Gaduš, J., Kliský, Ľ., Činovská, Ľ., & Dudák, A. (2012). Verification and optimization of biogas production from amaranth phytomass in fermentors bps kolíňany. In Markoš, J. (Ed.), *Proceedings of the thirty-ninth international conference of Slovak society of chemical engineering* (pp. 1324–1332). Tatranske Matliare, Slovakia.

Theegala, C. S., & Midgett, J. S. (2012). Hydrothermal liquefaction of separated dairy manure for production of bio-oils with simultaneous waste treatment. *Bioresource Technology, 107*, 456–463.

Williams, D. (2019). A floating device created to clean up plastic from the ocean is finally doing its job, organizers say, *CNN Business,* https://edition.cnn.com/2019/10/02/tech/ocean-cleanup-catching-plastic-scn-trnd/index.html.

CHAPTER 3

Source reduction and waste minimization in electrical and electronics industry

3.1 Introduction

Electrical and electronics industry denotes industrial setting that involves manufacturing, production, and distribution of electrical and electronic appliances. Despite being considered as commonly interchangeable terms, there are several differences in between electrical and electronic appliances. Most importantly, electrical devices can convert electrical energy into other energies whereas electronic devices cannot do so. Instead, electronic devices regulate the flow of electrons to perform corresponding device mechanisms. Though there are significant share of interrelations in between electrical and electronic devices, at the basic concept of electrical and electronic devices, there is a primary difference that differentiate electrical and electronic appliances. Electrical devices are mostly used to produce and utilize different forms of energy such as heat, sound, light, and so on. So, if the device is capable of energy conversion from electric energy to the other forms, it is an electrical device. On the contrary, if the device is not focused to convert electrical energy to other energies, instead, if it uses electron flow to maintain the device-related activities, then it can be considered as an electronic device. For example, a basic example of electrical device might be fan or light, since these appliances convert electric energy into other forms of energy to produce utilities; whereas transistor can be regarded as an electronic device, since it doesn't deal with energy conversion mechanism, instead electronic devices such as transistor regulate flow of electron, thus contribute to system control mechanism mostly. However, in recent days, the concept of electrical and electronic appliances is often integrated to assure more efficient service producing electrical and electronic appliances.

Electrical and electronics devices are the primary production of electrical and electronics industries. Electrical industry denotes industries that are involved with electricity or electrical appliances' generation, supply, distribution, and sale. On the other hand, electronics industry denotes

Source Reduction and Waste Minimization
http://dx.doi.org/10.1016/B978-0-12-824320-6.00003-4

automated and semi-automated industries that involve production, transmission, distribution, and sale of electronics appliances like television, computer, transistor, semiconductor, etc. However, the wide use of electrical and electronic industry produced appliances denotes how important their role is in terms of the technological and technical advancement. The contribution of such industries can be easily understood in recent days since we all are dependent on different electrical and electronic appliances in many ways. Therefore, the expansion of electrical and electronic industry is highly recommended to keep pace with the modern era of technological advancement. Being the world's fastest growing industry, electrical and electronics industry require incorporation of a variety of materials and methodologies (Babu et al., 2007). Being such a vast industrial setting, electrical and electronics industry often require involvement of good share of resources and manpower as well. From 2013 to 2017, electrical and electronics industry experienced a compound annual growth rate (CAGR) of 4.9% that is anticipated to rise at least over 6% between 2017 and 2024 (Zion Market Research, 2018; Market Research, 2018). This depicts how fast pace electrical and electronics industry can be.

Being a vast industry, electrical and electronics industry is also responsible to produce good share of waste production. All the varieties among stakeholders and complexities of corresponding methodologies associated with electrical and electronics industries make it tougher to coordinate and regulate the whole waste management system of the electrical and electronics industry effectively. When appliances that use electric power supply come to an end-of-life, they are categorized as waste from electronic and electrical equipment (WEEE). Three categories of waste cover 90% of world's WEEE of which 42% are large household electrical and electronic appliances, 34% are ICT-related appliances, and 14% are consumers' electronics (Reena et al., 2011). Such WEEEs are commonly known as e-waste and mostly generated from electrical and electronics industries during manufacturing or produced at community level after end use. In current context, the amount of produced E-waste is so huge that it is found that more than 4000 tons of e-waste is generated per hour with an annual growth of 4% (Ari, 2016; Srivastava and Pathak, 2020). Increased availability of electricity around the nook and cranny has made it easier to access a variety of electrical and electronic appliances. Therefore, considering the growing nature of this industry, urgency to take suitable measures in minimizing electrical and electronics industry generated waste has increased.

Due to the involvement of several complex processes and incorporation of hazardous chemicals during these processes causes electrical and electronics industry to produce highly toxic and harmful waste. Different forms of e-waste, produced during manufacturing or after end use, comprise more than thousands of toxic substances (Shamim et al., 2015). The situation is even more complex nowadays due to the decreased lifespan of e-appliances. For example, since lifespan of computers in 2015 reduced to half of the average lifespan of computers used in 1992, computer-generated e-waste has been double in this time duration (Puckett and Smith, 2002). Additionally, e-waste treatment facilities are still inadequate. Even the e-waste treatment sites are quite ineffective as they are found to disrupt ecological system and human health condition largely (Kiddee et al., 2020). Along with these, end-user produced e-waste are hardly treated or reduced; especially in the low income countries. Even if the industries take the role of minimizing the industrial e-waste produced during manufacturing, incapacity to treat end-user produced e-waste increases the net generation of e-waste. Besides, occupational hazard from such treatment plants is also quite significant that can be minimized if the electronics and electrical industries take some initiatives in reducing the e-waste by modifying their product generation mechanism (Kiddee et al., 2020). Thus, the urgency of waste minimization at industrial source-level is highly recommended in order to overcome the potential hazards.

This chapter discusses some of the major types of waste produced in the electrical and electronic industries. Additionally, it covers suitable waste source reduction mechanisms in such industries.

3.2 Commonly produced waste in the electrical and electronics industries: source and biogeochemical hazards

As discussed earlier, electrical and electronics industries produce a variety of highly toxic waste during generation and even after end use. Considering many different types of commonly produced e-waste, categorizing these wastes under suitable classes is necessary, since categorizing the waste under organized classes will help to assure better treatment and better management of the e-waste. However, commonly produced wastes in the electrical and electronics industries are often categorized into metallic waste and organic waste. Though there are many more different classes of waste generated in such industries, these are the two major industrial e-waste that require source-level minimization or immediate postgeneration level minimization. In addition to these, some amount of ceramics, glass, alloys, stabilizing agents, and pigments are also produced as industrial waste which can also be reduced, reuse, recycled, or treated under several mechanisms.

3.2.1 Metallic waste

Metallic e-waste denotes industrial wastes that are metals or metal ions. Metallic waste is the major category of commonly produced waste in the electrical and electronics industry, since most of the electronic or electrical appliances require incorporation of several metallic components. There are huge variety of metallic waste produced in electrical and electronics industry. Table 3.1 covers a variety of metallic waste produced in electrical and electronics industries along with their potential e-sources and hazards.

Table 3.1 Metallic e-waste sources and hazards (Srivastava and Pathak, 2020; Korla and Mitra, 2020; Chen et al., 2011; Ilankoon et al., 2018; Pathak and Srivastava, 2017).

Pollutant	E-sources	Hazards
Aluminum	Electricity transporter	• Acidification of environmental components • Inhibits metabolic enzymes • Impairs root growth at high concentration
Antimony	Diodes, semiconductors, heat resistant costing of e-appliances	• Groundwater contaminant • Soil pollutant • Kills microbes and small animals • Eye or skin irritation • Lungs, heart, or stomach problems
Arsenic	Light emitting diodes, GaAs devices	• Groundwater contamination leads to arsenicoses, keratosis, hypertension, pigmentation, ischemic heart diseases, edema, anemia, etc. • Causes oxidative stress and nutritional deficiency when converted to arsenite in plant body.
Barium	Cathode-ray monitor, lamps	• Bioaccumulation • Reduced blood minerals • Cardiac arrest • Paralyze organs
Beryllium	Ceramic components of electronics, X-ray machines, computers, power supply system	• Liver/kidney/heart/lymphatic damage • Cause beryllium sensitization for long-term exposure • Interferes mineral uptake

Table 3.1 Metallic e-waste sources and hazards (Srivastava and Pathak, 2020; Korla and Mitra, 2020; Chen et al., 2011; Ilankoon et al., 2018; Pathak and Srivastava, 2017). (*Cont.*)

Pollutant	E-sources	Hazards
Cadmium	Rechargeable computer batteries, semiconductor chips, photocopying toner, mobile phone, infrared detector, Ni-Cd batteries, circuit board	• Kidney/bones/reproductive/respiratory damage • Bioaccumulation • Highly toxic to environment
Chromium	Coating components, floppy disks, data tapes	• Common and highly toxic water pollutant • Carcinogenic • Damage reproductive system/endocrine system
Cobalt	Magnets, batteries, superalloy formation	• Overexposure leads to asthma, pneumonia, visual impairment, dermatitis, thyroid malfunction, and other health problems • Carcinogenic
Copper	Wire, circuit board	• Bioavailability determines hazards • Higher concentration is toxic as it forms complex metallothionein • Enzyme inhibition • Initiates peroxidative chain reaction forming hydroxyl radical that degrades membrane lipids • Extremely toxic for aquatic animals
Gallium	Electric field, semiconductors, metal alloys, electric diodes	• Enhance effects of arsenic contamination • Water pollutant • Radioactive under certain conditions • Throat irritation, breathing issue, and chest pain
Germanium	Metal alloy, semiconductor, diodes, transistor	• Lung, kidney, nervous damage • Organ dysfunction • Water and air pollutant • Germanium containing pollutants can travel a long distance and can affect many individuals
Iron	Motor, generator, transformer, magnet, machine design	• High tissue concentration causes physiological imbalance • Iron overload causes oxidative stress

(*Continued*)

Table 3.1 Metallic e-waste sources and hazards (Srivastava and Pathak, 2020; Korla and Mitra, 2020; Chen et al., 2011; Ilankoon et al., 2018; Pathak and Srivastava, 2017). (*Cont.*)

Pollutant	E-sources	Hazards
Lead	Batteries, lamps, PC monitors,	• Intellectual impairment • Kidney/reproductive/respiratory damage • Bone deformation • Soil infertility
Lithium	Lithium-ion battery, mobile batteries, electrical system of aircraft	• Interfere in metabolic processes both in plants and animals, as it is a small ion
Manganese	Steel and alloy, resistors, heating elements	• Excessive exposure leads to osteoporosis and neurodegenerative diseases
Mercury	All sorts of flat screen lighting devices, circuit system, thermostats, monitors, cold cathode fluorescent lamps	• Bioaccumulation and biomagnification • Nervous system/kidney damage • Anemia
Molybdenum	Wire, resistor, element for material creep/expansion, heat tolerant elements of e-appliances	• Cause infertility among animal species • Reduce growth rate and organ maturation • Exposure over the safer limit significantly reduces agricultural production • Water contaminant and leads to fish habitat destruction • Molybdenum containing irrigated water disrupts plant growth
Nickel	Batteries, electric vehicles, metal alloys, stainless steel, corrosion-protective agent	• Skin irritation/itching/allergy • Long time exposure cause cancer damaging, lungs, nose, synovial, and throat function • Affects fertility • Cause hair or fur loss • Regarded as immunotoxic, genetoxic, and neurotoxic for all animals
Selenium	Solar cells, printers, electric data sheets	• Affect endocrine and gastrointestinal functions negatively • Impairs cellular functions • Immunotoxic and hepatotoxic

Table 3.1 Metallic e-waste sources and hazards (Srivastava and Pathak, 2020; Korla and Mitra, 2020; Chen et al., 2011; Ilankoon et al., 2018; Pathak and Srivastava, 2017). (*Cont.*)

Pollutant	E-sources	Hazards
Silicon	Semiconductor, insulator, dynamo, transformer plates, machine tools	• Despite being less toxic, occupational exposure or inhalation leads to rheumatoid arthritis, silicosis, sarcoidosis, and other diseases • Though it helps to release heavy metal stress in plant, excessive exposure suppress plant growth, and plants' antioxidant system
Silver	Films, conventional switches, printed circuit boards, silver-based inks, electrical pathways	• Eye or skin pigmentation • Irritation in sensory organs • Liver and kidney damage • Plant growth is impaired
Tin	Batteries, alloys, coating agent	• Exposure can lead to stomachaches, damage of internal organs (liver, kidney, nervous system, etc.) or anemia • Disrupts aquatic environment • Kills phytoplankton and useful microbes
Tungsten	Filaments of incandescent lamps, vacuum tubes, carbide bits	• Seedling, root, and shoot growth is hindered • Structural malfunctions, aberration of cell cycle, and immune and cytoskeletal disruption • Hamper gene expression
Zinc	Batteries, bodies of electrical or electronics parts	• Dizziness • Fatigue

3.2.2 Organic waste

In addition to different metallic pollutants, electrical and electronic industry is also found to produce good amount of organic pollutants which should also be treated and minimized at the source level. Table 3.2 explores some of the major organic waste produced in electrical and electronics industries along with their potential e-sources and hazards.

Table 3.2 Organic e-waste sources and hazards (Srivastava and Pathak, 2020; Korla and Mitra, 2020; Kiddee et al., 2020; Chen et al., 2011; Ilankoon et al., 2018; Pathak and Srivastava, 2017).

Pollutants	E-sources	Hazards
POPs	Circuit boards, plastic components of computers, fluorescent light, fans, motor, connectors, lubricants in e-appliances, mobile phones	• Resistant to break down in environment/ bioaccumulation • Air pollution • Neurotoxicity • Memory function interference • Hormonal imbalance for long-term exposure
PVC (polyvinylchloride)	Insulators, wires, and cables	• Improper management leads to toxic furans and chlorinated dioxins generation • Even low concentration is harmful • Persistent • Pulmonary damage
PCB (polychlorinated biphenyls)	Capacitors, transformers, dielectric fluids, lubricants, electric motors	• Affect nervous, endocrine, cognitive, and immune system • Skin conditions degradation • Soil and water pollutant
DL-PCB (dioxin-like polychlorinated biphenyls)	Dielectric fluid, lubricant, coolants, transformers, capacitors, generators, fluorescent lighting	• Affect fetal growth and infant development • Organ dysfunction for long term exposure
BFR (brominated flame retardant)	Industrial chemical, process component	• Neurobehavioral damage causing lower IQ, memory problem, retarded physical and mental growth • Morphological impairment • Bioaccumulation occurs and enters in food chain
PAH (polyaromatic hydrocarbons)	Combustion emission from electronic products	• Long term exposure leads to skin inflammation, organ damage, and other health consequences • Not safe for most of the species; especially for aquatic environment

Table 3.2 Organic e-waste sources and hazards (Srivastava and Pathak, 2020; Korla and Mitra, 2020; Kiddee et al., 2020; Chen et al., 2011; Ilankoon et al., 2018; Pathak and Srivastava, 2017). (*Cont.*)

Pollutants	E-sources	Hazards
PBDEs (polybrominated diphenyl ethers)	Heat and fire retardant objects	• Leads to several different types of toxicity such as neurotoxicity, immunotoxicity, reproductive toxicity, and others
PCDD/Fs (polychlorinated dibenzodioxinsand dibenzofurans)	Combustion by-product, hydrocarbon receptors	• May cause breathing problem at high concentration

3.3 Source-level waste minimization initiatives by the electrical and electronics industries

E-waste minimization at source level is highly associated with process designing for recycling and resource recovery. In high income countries, industrial initiatives to ensure source-level e-waste minimization are quite adequate where low income developing countries often struggle to modify industrial processes rather such countries often adapt with informal and decentralized e-waste minimization. Most of the electrical and electronics industries are found to dispose a huge share of their waste in landfill sites. Landfill disposal of e-waste is not a very efficient waste minimization strategy. Toxic substances of e-waste are transferred by water and soil tension to percolate and leachate. Such leaching contains high level of hazardous, dissolved, and suspended heavy metals and organic compounds of e-waste (Kiddee et al., 2020). The leaching capacity of metals and organic pollutants vary in from component to component. Even decomposition of e-waste components are found to produce extremely toxic gaseous hazards that has environment and health degradation potentials. Thus, traditional inefficient waste source reduction mechanisms are found to be environmentally unsustainable and detrimental for health.

However, modern innovations can be applied in such contexts to ensure more efficient waste minimization at source that includes process modification strategies, recycling approach, and policy interventions. Fig. 3.1 illustrates some of the major waste source reduction strategies for electrical and electronics industries.

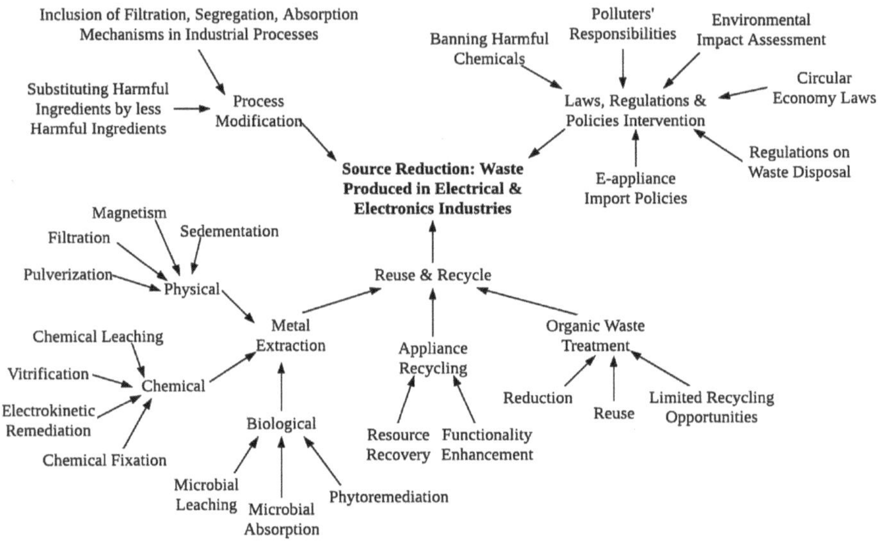

Figure 3.1 ***Overview on waste source reduction strategies in electrical and electronics industries.***

3.3.1 Process modification

Process modification is regarded as the core waste source reduction mechanism, since it helps to reduce waste before its generation. Several modifications in the processes used in electrical and electronics industries not only reduces generation of waste but also helps to shift from more toxic waste to less toxic waste generation. Process modification may occur in different ways such as product substitution, clearer technology inclusion, technological advancement, efficiency enhancement, and others. For efficient process modification, initially the harmful products should be substituted by less harmful ingredients. For instance, increasing use of lithium-ion battery is ensured, as it is more environmental friendly and less toxic for human health comparing with other conventional batteries (Di Lecce et al., 2017). Therefore, a significant shift has occurred from conventional batteries to lithium-ion batteries. Likewise, along with substitution of ingredients, processes should also undergo modifications as per waste minimization guidelines. More waste generating systems should be substituted by less waste generating system. Additionally, filtration, separation, absorption mechanisms should be introduced in different steps of industrials processes for gaseous and liquid pollution prevention (Sutherland, 2008). Along with these, the efficiency of the existing facilities should also be increased so that there is less amount of by-product produced as waste through efficient system mechanism. Unless such different process modification initiatives are introduced, incorporation of other source-level or immediate post waste minimization strategies will be of no use due to the ever-rising amount of highly toxic amount of waste produced from electrical and electronics industries. This is why, integration of suitable waste minimization mechanisms are needed to be incorporated at several steps of industrial processes.

3.3.2 Recycling

Recycling waste is one of the core ways to minimize source-level waste generation. In specific context of electrical and electronics industry, recycling mechanisms rely on adoption of several tools with an aim to extract metals, recycle organic waste, reuse produced appliances, and so on. Though post-production level recycling of WEEE is often noticed, preproduction level waste minimization through recycling is not widely practiced (Tsai, 2020). It is comparatively less popular due to industrial reluctance, technological limitation, and policy lacking. Such hindrances can be significantly overcome by ensuring producers' contribution in recycling produced waste at preproduction level; especially by treating industrial sludge, reusing useful components of waste and by installing suitable industrial setup for recycling facilities (Ayvaz and Görener, 2020). Though it may seem cost inefficient and troublesome, the useful products and sustainability gained from recycling will trade off the setup cost (Kumar et al., 2020). However, there are evidences of limited-scale waste source reduction through recycling in electrical and electronics industry. Some of such strategies and mechanisms are discussed later.

3.3.2.1 Metal extraction from e-waste

As discussed earlier, production of electrical and electronic appliances require a good number of metallic products that are released along with industrial sludge in our environment, considering the potential hazards that such metallic waste may cause, it is urgent to recycle and reuse these metallic waste. Adoption of multiple strategies will recover a variety of metals in good amount. To depict the potentials of metal recovery from e-waste producing electrical and electronic industries, the overall gold production scheme can be noticed. More than one-tenth of the total gold production in 2013 came from e-waste based metal recovery processes (USGS, 2014). Likewise, a good share of other valuable metallic production came out of electrical and electronics waste treatment. Therefore, if industrialists are encouraged to adopt such initiatives, they will also be largely benefitted.

Metal extraction mechanisms can be as simple as physical segregation to as complex as bioremediation. Considerable amount of reusable metals can be isolated from e-waste introducing preliminary segregation mechanism. Isolated metals like Cu, Ag, Pd, etc. can be used to make circuit boards, batteries, etc. Isolated Pd and Pt can be used for catalytic converter (Korla and Mitra, 2020). Such metal extraction processes are more efficient at source level, as the hazards are significantly minimized, as exposure to environment and health system is strictly prohibited. Suitable source reduction strategies may require incorporation of physical, chemical, biological, and biochemical mechanisms based on the industrial contexts. Some of such strategies are introduced in this chapter.

1. **Physical treatment for metal extraction**
   Physical process can be adopted in electrical and electronics industries that will isolate metallic pollutants from waste sludge. Physical process denotes combination of mechanisms that do not alter the chemical formulation of the waste and do not require intervention of any biological or chemical agent. Such processes often require a landfill site for waste disposal where series of mechanisms such as soil substitution, sweeping, and land importation take place (Zhang et al., 2004). Such traditional physical mechanisms for metal recovery are not well-implemented yet, as it is only applicable in small scale and it is not yet a very efficient mechanism. Currently, in addition to traditional physical metal extraction strategies, a variety of tools are also introduced to increase efficiency. It includes magnate-based separation, shredding, pulverization, and many more approaches (Ayangbenro and Babalola, 2017). Such physical treatment mechanisms are often regarded as primary metal extraction mechanism before implementing other advanced bio-chemical processes.
2. **Chemical treatment for metal extraction**
   Chemical treatment of waste at a source level requires inclusion of sophisticated methodologies and reagents in order to extract metallic components from waste components. Unlike physical metal extraction process, in case of chemical remediation, chemical alteration of the components occurs. Chemical remediation of waste often requires certain atmospheric conditions (temperature, pressure, etc.) in addition to specific chemical reagents. Some of the commonly used chemical metal extraction strategies involve vitrification, chemical leaching, electrokinetic remediation, chemical fixation, and many more.

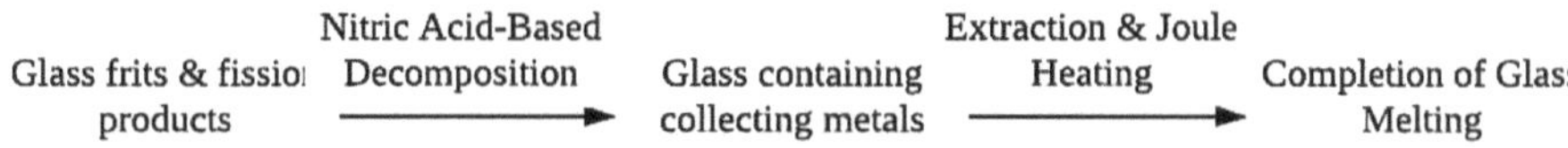

Figure 3.2 ***Mechanism of vitrification process (Uruga et al., 2007).***

a. Vitrification
Vitrification is the technology that uses soil of extremely high temperature (1400°C–2000°C) to decompose or volatize matters (Dhaliwal et al., 2020). In this process, fossil fuels, microwave system, plasma, etc. are used for energy supply. Electrodes are inserted in the waste in case of direct vitrification so that the metallic waste can accumulate in the electrodes and in case of indirect vitrification sometimes solvent medium is used for electrode insertion (Goswami and Das, 2015). At the end of the process, the accumulated metals can be isolated from the electrodes. However, despite being a very efficient technology, this is not quite popular due to its high implementation cost. Fig. 3.2 illustrates some of the basic steps associated with vitrification process.

b. Chemical leaching
Chemical leaching is the process that uses chemical reagents, fluids and gaseous substances to eliminate contaminants from a soil or water system. In this process, metallic components are transferred to liquid phase by precipitation, chelation, adsorption, ion exchange, or by other mechanisms (Dhaliwal et al., 2020). Based on the types of contaminants present in the system, acids of different concentration are used for suitable chemical leaching (Tokunaga and Hakuta, 2002). Despite being a very efficient metal extraction system, effects of chemical leaching are not much environment friendly, as it uses a lot of chemicals that may affect environmental components negatively.
Reactions and reagents for chemical leaching vary based on the targeted metal to be recovered. For instance, potential copper and silver recovery mechanisms for electrical and electronics industrial waste can be as follow (Serpe et al., 2015).
Copper recovery through chemical leaching:

$$\mathrm{Cu^{2+}(aq) + 4NH_3(aq) \rightleftharpoons \left[Cu\left(NH_3\right)_4\right]^{2+}(aq)}$$

$$\mathrm{\left[Cu\left(NH_3\right)_4\right]^{2+}(aq) + 2e^- \rightleftharpoons Cu(s) + 4NH_3(aq)}$$

Silver recover through chemical leaching:

$$AgI(s) \rightleftharpoons Ag^{+}(aq) + I^{-}(aq)$$

$$AgI(s) + 1e^{-} \rightleftharpoons Ag(s) + I^{-}(aq)$$

$$Ag^{+}(aq) + 2NH_{3}(aq) \rightleftharpoons \left[Ag\left(NH_{3}\right)_{2}\right]^{+}(aq)$$

$$\left[Ag\left(NH_{3}\right)_{2}\right]^{2+}(aq) + 1e^{-} \rightleftharpoons Ag(s) + 2NH_{3}(aq)$$

c. Electrokinetic remediation

Electrokinetic remediation is the process that uses voltage difference of electric field gradient. In this process, charged components of waste are separated using electrokinetic force. This process is very efficient in low permeable waste system. Besides, this is a cost effective and less complex system (Dhaliwal et al., 2020). This process is mostly regulated by electromigration and electroosmotic flow mechanism. However, unlike many other chemical remediation strategies, this process is not much efficient in terms of its metal recovery potentials due to the difficulties associated with soil pH maintenance (Fasani et al., 2017). Besides, large scale implementation of such approaches in the industrial scale is yet a challenge.

d. Chemical fixation

Chemical fixation denotes addition of chemical reagents in waste materials so that slightly insoluble materials can be formed limiting contaminant movement in the system (Dhaliwal et al., 2020). Therefore, chemical fixation is often regarded as a stabilizer mechanism for metallic pollutants. Suitable chemical reagent selection is dependent on the types of pollutants present in the waste cluster. For example, finely grounded crystalline $Ca_{10}(PO_4).6H_2O$ is used to immobilize metal phosphate contaminants (Hodson et al., 2000).

3. Biological treatment for metal extraction

Metal recovery from e-waste using biological processes is another emerging area of e-waste minimization. Biological metal extraction mechanisms can be implemented using either microbial remediation or phytoremediation mechanisms. Both of these processes are well-implemented at an industrial scale.

a. Microbial remediation

As microbial remediation combines principles of biotechnology and metallurgy, such implications are part of biometallurgy. This concept is based on microbial and metallic interactions. As the metals are found mostly in the form of minerals, several bacteria and fungi have the enzymatic or bio-absorptive properties for metallic component isolation from minerals. Bacterial and fungal metal extraction mechanisms vary from each other, though the basic principle incorporates microbial ore leaching or microbial absorption mechanism (Dhaliwal et al., 2020). In case of ore-leaching mechanism, suitable bacteria/fungi enzymes attach to waste cluster and initiate direct or indirect microbial leaching. For example, the bacteria, *T. ferrooxidans* can cause an ATP coupled electron shift from iron to oxygen. That helps to isolate iron ore from complex iron compounds. In this regard, the general reaction of iron recovery is as shown (Benzal et al., 2020).

$$4Fe^{2+} + 4H^{+} + O_2 \xrightarrow{\text{Biological activity}} 4Fe^{3+} + 2H_2O$$

Likewise, other valuable metals can also be isolated using microbial leaching mechanism.

In this regard, a basic workflow for microbial leaching can be as shown in Fig. 3.3.

In addition to microbial leaching, microbial bio-absorption can be another leading way through which useful metals are up-taken by suitable microorganisms. Such microbial electrometallurgy mechanism requires suitable microorganisms that regulated the process through microbial surface and metallic interaction in suitable environmental setting (Dominguez-Benetton et al., 2018). Such

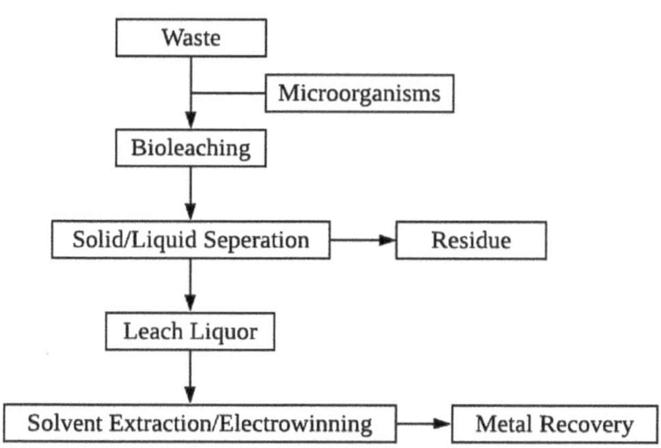

Figure 3.3 ***Basic workflow for microbial leaching (Sutar and Awasare, 2015).***

absorption processes are often chemical assisted that enhances or reduces absorption properties of the corresponding microorganisms (Sheel and Pant, 2018). Unlike microbial leaching mechanism, microbial adsorption is not dependent on enzymatic activities. Rather surface properties denote microbial absorption potentials. The means of absorption can be of various types; that is, bioaccumulation, electronic interaction, and so on. Absorbed metals can be isolated from microbial cluster later.

As microorganisms are metal specific in terms of their corresponding metal extraction capacity, Dhaliwal et al. (2020) carefully observed microbial metal extraction potentials from multiple researches and suggested suitable microbes for specific metallic isolation. Dhaliwal et al. (2020) explores the role of different microorganisms for successful metal extraction which can be applicable for source-level waste minimization in the electrical and electronics industries.

b. Phytoremediation

In addition to microbial treatment, phytoremediation is another popular biological treatment mechanism for metal extraction from industrial waste. Phytoremediation is a plant-assisted process that reduces toxic materials in environment by absorbing them from soil using suitable plants. In this regard, metal uptake and bioavailability is regulated by addition of chelating agents, such as, amendments, acid/base, fertilizers and so on (Dhaliwal et al., 2020). Some specific plant species are well noted to implement phytoremediation which are known as hyper-accumulators. These plant species can tolerate high metallic concentration without causing disturbance of its growth rate and biomass generation (Memon and Schroder, 2009). Dhaliwal et al. (2020) and Sumiahadi and Acar (2018) detects some of the leading hyper accumulator plants along with corresponding metallic classes that it can accumulate. This can be applicable for metallic waste producing industries such as electrical and electronics industries to minimize good amount of metallic waste from the disposal sites.

Electrical and electronics industries can plant such plants in their sludge disposal sites to ensure metal absorption. However, using this process metal recovery cannot be ensured always, as metal isolation potentials from the plants are a highly complicated process. Instead, this is more of a source-level metallic waste isolation mechanism. Considering the implementation simplicity and cost-effectivity of the process, phytoremediation is a very popular waste reduction process in several contexts; especially in case of smaller industrial setting.

3.3.2.2 Organic e-waste reduction, reuse, and recycling

Unlike metallic waste materials, organic waste produced in electrical and electronics industry often do have much significant usage for which extraction of organic waste through complex physical, chemical, and biological mechanisms are not much worthy to be implemented. Therefore, organic electrical and electronics industrial waste reduction and reusing is prioritized over recycling organic waste. First, unlike other industrial waste minimization strategies, organic waste reduction is often accompanied by incineration or burning of organic waste (Pérez-Martínez et al., 2019). Additionally, as there is a tendency to maximize resource efficiency, a significant reduction in organic matter utilization has happened recently comparing with former inefficient practices, which has also contributed in organic waste reduction. Second, to ensure organic waste reuse, use of organic matters that can used multiple times are being enhanced. For instance, heating plastics like PVC are very durable to be reused (Korla and Mitra, 2020). Besides, organic wastes are also used as energy source. Electrical and electronics industry can consider their generated share of organic waste as energy source. Thus, reduction and reuse of organic waste will both add value to the industrial benefits and waste source reduction.

Recycling of some share of organic waste produced in electrical and electronics industry is also needed. For instance, PVC used in electrical and electronic appliances synthesis can be transformed into other PVC goods used at an industrial scale. However, most of the organic wastes produced in the industries are highly toxic for both environment and human health. Therefore, considering the recycling related limitations of organic waste, reuse and reduction of electrical and electronics industry produced organic waste should be focused.

3.3.2.3 Appliance recycling

Electrical and electronics appliance recycling is comparatively more popular and widely implemented. Though it is not a source-level waste minimization mechanism, it is urgent to consider appliances recycling seriously, as such appliances are one of the leading causes of large scale pollutant emission. Most of such appliances recycling initiatives are regulated by informal sectors in developing countries (Chi et al., 2011). Thus a significant share of resources become waste due to limited scope of resource and service recovery opportunities. In addition to it, such informal initiatives are practiced in a highly contagious and unsafe environment causing occupational hazards to the people engaged in such activities (Annamalai, 2015). Therefore, a sophisticated formal setting for such appliances recycling is needed. Many developed countries such as Switzerland, United Kingdom, and Denmark have already introduced formal government and private sector regulated initiatives to reuse and recycle electrical and electronic appliances (Kang and Schoenung, 2005; Parajuly et al., 2017; de Oliveira Neto et al., 2017). But as such initiatives are only concentrated for high income countries, global burden caused by e-waste is not deducting significantly.

In this regard, the role of such appliance producers is important. It is because, being the producers, the corresponding industries already have a suitable industrial setting that need to be expanded slightly for appliance recycling facilities. In this process of appliance recycling, the electrical and electronics industries can either recover useful resources or can enhance functionality of the appliances. This will help the industries to gain monetary benefits and sustainability in the long run.

3.3.3 Laws, regulations, and policies intervention

Only technological adoption is not enough to ensure sustainable waste source reduction in electrical and electronics industries. Rather interventions of policies, laws, and regulations simultaneously are also needed. In this regard, developed countries are quite advanced comparing with the developing countries. Therefore, in addition to technological adoption, developed countries could tackle the burden of e-waste generation quite successfully. For instance, 23% of e-waste are produced in developed countries but sent to developing countries (Balde et al., 2015). It is one of the major tactics of developed countries to avoid the burden of e-waste production. Developed countries like Switzerland, despite being one of the biggest producers of electrical and electronic waste, recycle 75% of their produced e-waste (Hischier et al., 2005). On the other hand, in developing countries like India, only 1%–2% e-wastes are effectively recycled and mostly women and children participate in this recycling process without proper safety (Korla and Mitra, 2020). This also indicates the developing countries' failure to introduce suitable policy instruments and legislation in regulating indirectly imported e-waste which is a major reason behind their failure to reduce their e-waste (Kiddee et al., 2020). Apart from these, there are many policy-regulated approaches that are hoped to help largely in waste source reduction mechanism (Table 3.3).

Table 3.3 Laws, regulations, and policies to minimize source-level e-waste (Srivastava and Pathak, 2020; Ministry of Environmental Protection (MEP), 2003; Yang et al., 2008; Streicher and Yang, 2007).

Regulation strategies	Contents
Banning harmful chemicals	Harmful chemicals are banned to be utilized in e-industries imposing policy interventions of government or international agencies which is a source-level e-waste minimization strategy; that is, Chinese government's ban on importation of seven categories of waste in 2000.
Polluters responsibilities to minimize pollution	Under this regulation, the polluters of all levels (manufacturer, consumers, etc.) are responsible to take or contribute in initiatives to ensure reduction in e-waste. For instance, Brazil's "Law of Garbage" states everyone's responsibilities in minimizing their generated waste. In this regard, the contribution might be taxation of users, process modification of manufacturers, and many more activities.
Environment-friendly e-waste minimization	Such strategies of the regulating bodies encourage industrialists for clean manufacturing processes and eco-friendly process designing in exchange of incentives.
Regulations on industrial waste disposal	E-industries are obliged to follow labelled safer concentration of components in e-waste while disposing.
E-appliances import policy	A strategic system that limits importing e-appliances with an aim to reduce the country's overall e-waste generation. For instance, Import policy of India in 2005 put restrictions in import of certain pollutant containing e-appliances.
Circular economy laws	Regenerative process to specify provisions for 3R approach in different stages of product production, utilization, and disposal.

However, a balanced well-regulated system in association with technological innovations and policy implementation is needed to ensure sustainable waste source reduction scheme in electrical and electronics industries.

Case Study 4: Waste to Resource Concept Implementation through Metal Recovery Mechanisms Introduced by External Firms

Since many of the industries often find it difficult to install suitable mechanisms for successful source-level waste minimization mechanisms, they can ask for consultancy or technical support from other expert agencies. In this regard, the example of GCL group can be a good example for many countries; especially the ones who are facing hardships to design suitable resource recovery initiatives as a part of their waste minimization strategies. GCL, established in 2014, is an international recycling and trading group, founded in Istanbul who supports electrical and electronics industries to set up suitable recycling and resource recovery. They offers a two way mechanism that not only reduce hazardous waste through successful implementation of technologies, at the same time, they also guide to recover good amount of resources from the installed facilities.

This is a quite encouraging example since such an enthusiastic group coming from Turkey that is not even a heavily industrialized country took the initiative of source-level waste minimization that should have been done by many other richer or heavily industrialized countries. What makes GCL more extraordinary is their role of waste minimization initiative for the electrical and electronics industries through modification of the whole supply chain mechanism. Despite being a group run by only 300 employees, GCL has recycled 300 tons of catalysts, recovered 450 tons of copper, 2 tons silver, and 400 kg of golds annually from only electronic waste (GCL, 2020). This fact illustrates that if this group didn't take the initiative for effective source-level waste minimization, a vast amount of resource would have been wasted. In this regard, such national or international level initiatives should be encouraged and promoted for the industrial as well as for the country's welfare so that maximum outputs can be achieved from limited resources.

References

Annamalai, J. (2015). Occupational health hazards related to informal recycling of E-waste in India: An overview. *Indian Journal of Occupational and Environmental Medicine, 19*(1), 61.

Ari, V. (2016). A review of technology of metal recovery from electronic waste. In *E-waste in transition—from pollution to resource*. IntechOpen.

Ayangbenro, A. S., & Babalola, O. O. (2017). A new strategy for heavy metal polluted environments: a review of microbial biosorbents. *International Journal of Environmental Research and Public Health, 14*(1), 94.

Ayvaz, B., & Görener, A. (2020). Reverse logistics in the electronics waste industry. In *Waste management: concepts, methodologies, tools, and applications* (pp. 1664–1680). IGI Global.

Babu, B. R., Parande, A. K., & Basha, C. A. (2007). Electrical and electronic waste: a global environmental problem. *Waste Management & Research, 25*(4), 307–318.

Balde, C.P., Wang, F., Kuehr, R., & Huisman, J. (2015). *The global e-waste monitor 2014: Quantities, flows and resources*. Bonn, Germany: United Nations University, IAS–SCYCLE.

Benzal, E., Cano, A., Solé, M., Lao-Luque, C., Gamisans, X., & Dorado, A. D. (2020). Copper recovery from PCBs by Acidithiobacillus ferrooxidans: toxicity of bioleached metals on biological activity. *Waste and Biomass Valorization, 11*, 1–10.

Chen, M., Zhang, F.S., & Zhu, J. (2011). Leaching toxicity of Pb and Ba containing in cathode ray tube glasses by SEP-TCLP. In *Recycling of electronic waste II: proceedings of the second symposium* (pp. 109–113), Vol. 1. Hoboken, NJ, USA: John Wiley & Sons, Inc.

Chi, X., Streicher-Porte, M., Wang, M.Y., & Reuter, M. A. (2011). Informal electronic waste recycling: a sector review with special focus on China. *Waste Management*, *31*(4), 731–742.

de Oliveira Neto, G. C., Correia, A. D. J. C., & Schroeder, A. M. (2017). Economic and environmental assessment of recycling and reuse of electronic waste: Multiple case studies in Brazil and Switzerland. *Resources, Conservation and Recycling*, *127*, 42–55.

Dhaliwal, S. S., Singh, J., Taneja, P. K., & Mandal, A. (2020). Remediation techniques for removal of heavy metals from the soil contaminated through different sources: a review. *Environmental Science and Pollution Research*, *27*(2), 1319–1333.

Di Lecce, D., Verrelli, R., & Hassoun, J. (2017). Lithium-ion batteries for sustainable energy storage: recent advances towards new cell configurations. *Green Chemistry*, *19*(15), 3442–3467.

Dominguez-Benetton, X., Varia, J. C., Pozo, G., Modin, O., Ter Heijne, A., Fransaer, J., et al. (2018). Metal recovery by microbial electro-metallurgy. *Progress in Materials Science*, *94*, 435–461.

Fasani, E., Manara, A., Martini, F., Furini, A., & Dal, C. G. (2017). The potential of genetic engineering of plants for the remediation of soils contaminated with heavy metals. *Plant Cell Environment*, *41*, 1201–1232.

GCL (2020). Who we are: GCL Group. *GCL*. Available from: http://www.gclcevre.com/en/gcl-group.

Goswami, S., & Das, S. (2015). A study on cadmium phytoremediation potential of Indian mustard, *Brassica juncea*. *International Journal of Phytoremediation*, *17*, 583–588.

Hischier, R., Wäger, P., & Gauglhofer, J. (2005). Does WEEE recycling make sense from an environmental perspective?: The environmental impacts of the Swiss take-back and recycling systems for waste electrical and electronic equipment (WEEE). *Environmental Impact Assessment Review*, *25*(5), 525–539.

Hodson, M. E., Valsami-Jones, E., & Cotter-Howells, J. D. (2000). Bone meal additions as a remediation treatment for metal contaminated soil. *Environment Science & Technology*, *34*, 3501–3507.

Ilankoon, I. M. S. K., Ghorbani, Y., Chong, M. N., Herath, G., Moyo, T., & Petersen, J. (2018). E-waste in the international context–A review of trade flows, regulations, hazards, waste management strategies and technologies for value recovery. *Waste Management*, *82*, 258–275.

Kang, H. Y., & Schoenung, J. M. (2005). Electronic waste recycling: A review of US infrastructure and technology options. *Resources, Conservation and Recycling*, *45*(4), 368–400.

Kiddee, P., Pradhan, J. K., Mandal, S., Biswas, J. K., & Sarkar, B. (2020). An overview of treatment technologies of E-waste. In *Handbook of electronic waste management* (pp. 1–18). Butterworth-Heinemann.

Korla, K., & Mitra, C. K. (2020). Biochemical hazards associated with unsafe disposal of electrical and electronic items. In *Handbook of electronic waste management* (pp. 55–80). Butterworth-Heinemann.

Kumar, A., Wasan, P., Luthra, S., & Dixit, G. (2020). Development of a framework for selecting a sustainable location of waste electrical and electronic equipment recycling plant in emerging economies. *Journal of Cleaner Production*, *277*, 122645.

Market Research (2018). Electrical & electronics retail global industry guide 2018–2022. *MarketLine*. Available from: https://www.marketresearch.com/MarketLine-v3883/Electrical-Electronics-Retail-Global-Guide-11564058/.

Memon, A. R., & Schroder, P. (2009). Implications of metal accumulation mechanisms to phytoremediation. *Environmental Science and Pollution Research*, *16*, 162–175.

Ministry of Environmental Protection (MEP) (2003). *The circular on strengthening the environmental management of waste electrical and electronic equipment*. Available from: http://www.mep.gov.cn/gkml/zj/wj/200910/t20091022 172222.htm (in Chinese).

Parajuly, K., Habib, K., & Liu, G. (2017). Waste electrical and electronic equipment (WEEE) in Denmark: Flows, quantities and management. *Resources, Conservation and Recycling, 123*, 85–92.

Pathak, P., & Srivastava, R. R. (2017). Assessment of legislation and practices for the sustainable management of waste electrical and electronic equipment in India. *Renewable and Sustainable Energy Reviews, 78*, 220–232.

Pérez-Martínez, S., Giro-Paloma, J., Maldonado-Alameda, A., Formosa, J., Queralt, I., & Chimenos, J. M. (2019). Characterisation and partition of valuable metals from WEEE in weathered municipal solid waste incineration bottom ash, with a view to recovering. *Journal of Cleaner Production, 218*, 61–68.

Puckett, J. & Smith, T. (2002). *Exporting harm the high-tech trashing of Asia.* San Jose, CA: Silicon Valley Toxics Coalition.

Reena, G., Sangita, & Verinder, K. (2011). Electronic Waste: A Case Study. *Research Journal of Chemical Sciences, 1*(9), 49–56.

Serpe, A., Rigoldi, A., Marras, C., Artizzu, F., Mercuri, M. L., & Deplano, P. (2015). Chameleon behaviour of iodine in recovering noble-metals from WEEE: towards sustainability and "zero" waste. *Green Chemistry, 17*(4), 2208–2216.

Shamim, A., Ali, Mursheda, K., & Rafiq, I. (2015). E-waste trading impact on public health and ecosystem services in developing countries. *International Journal of Water Resources, 5*, 112.

Sheel, A., & Pant, D. (2018). Recovery of gold from electronic waste using chemical assisted microbial biosorption (hybrid) technique. *Bioresource technology, 247*, 1189–1192.

Srivastava, R. R., & Pathak, P. (2020). Policy issues for efficient management of E-waste in developing countries. *Handbook of electronic waste management*, 81–99.

Streicher, P. M. & Yang, J. X. (2007). WEEE recycling in China: Present situation and main obstacles for improvement. In *Proceedings of the 2007 international symposium on electronics and the environment* (pp. 40–45). Orlando, FL, USA: IEEE.

Sumiahadi, A., & Acar, R. (2018). A review of phytoremediation technology: heavy metals uptake by plants. *Earth and Environmental Sciences, 142*, 12–23.

Sutar, A. S., & Awasare, S. D. (2015). Bioleaching of ore using chemolithotrophic microorganisms (Acidithiobacillus ferrooxidans). *International Journal of Research Environment Science & Technology, 5*(1), 14–21.

Sutherland, K. (2008). Air filtration: Using filtration to control industrial air pollution. *Filtration & Separation, 45*(9), 16–19.

Tokunaga, S., & Hakuta, T. (2002). Acid washing and stabilization of an artificial arsenic-contaminated soil. *Chemosphere, 46*, 31–38.

Tsai, W. T. (2020). Recycling waste electrical and electronic equipment (WEEE) and the management of its toxic substances in Taiwan—A case study. *Toxics, 8*(3), 48.

Uruga, K., Sawada, K., Arita, Y., Enokida, Y., & Yamamoto, I. (2007). Removal of platinum group metals contained in molten glass using copper. *Journal of Nuclear Science and Technology, 44*(7), 1024–1031.

USGS, 2014. *Mineral Commodity Summaries 2014.* Reston, VA: U.S. GeologicalSurvey (USGS).

Yang, J. X., Lu, B., & Xu, C. (2008). WEEE flow and mitigating measures in China. *Waste Management, 28*, 1589–1597.

Zhang, Y. F., Sheng, J. C., & Lu, Q.Y. (2004). Review on the soil remediation technologies. *Gansu Agriculture Science and Technology, 10*, 36–38.

Zion Market Research (2018). Consumer electronics market by product (smartphone, television, DVD players, refrigerators, washing machines, digital cameras, and hard disk drives) and by region (North America, Europe, Asia Pacific, Latin America, and the Middle East & Africa)—global industry perspective, comprehensive analysis and forecast, 2017–2024. *zionmarketresearch.com.* Available from: http://www.zionmarketresearch.com/report/consumer-electronics-market.

CHAPTER 4

Source reduction and waste minimization in healthcare industry

4.1 Introduction

Being the most crucial industry to ensure good health and wellbeing of the individuals, healthcare industry is one of the oldest industries that comprises up-to-date inclusions every day. Unlike many other industries, healthcare industry is highly fragmented because of having a lot of small players under centralized to decentralized system. Major segments of healthcare industry include healthcare services, medical equipment, pharmaceutical drugs, biologics, veterinary healthcare, and so on (Business Wire, 2019). Hospitals, biomedical laboratories, healthcare research centers, blood banks, nursing homes, autopsy centers, animal research laboratories, etc. are the common role players in healthcare industry (WHO, 2018). Among these stakeholders, healthcare service providers, having a compound annual growth rate (CAGR) of 9.3% in between 2014 and 2022, are the largest segment of the global healthcare market, as they are accounted for 79.4% of the total healthcare industry. Biologics and veterinary healthcare are found to be the fastest growing segments because of having a recent CAGR of 13.6% and 10.9%, respectively (Allen, 2020; Business Wire, 2019). However, the overall CAGR of the global healthcare industry is expected to be at least 5% from 2019 to 2022 (Allen, 2020). In this meantime, the value of the healthcare market will rise at nearly $11,908.9 billion in 2022 from the former value of $8,452 billion noted during 2018 (Business Wire, 2019). Considering the huge significance of the healthcare industry, country governments are trying to regulate it by introducing policies, laws, and regulations at different levels. Poor regulations may lead dreadful impacts in later terms.

Healthcare industries produce a variety of waste in different steps of industrial processes. These wastes are commonly regarded as "Healthcare Waste" or HCW. UNEP (2020) defines healthcare waste as "all the waste generated by healthcare facilities, medical laboratories and biomedical research facilities, as well as waste from minor or scattered sources."

Source Reduction and Waste Minimization
http://dx.doi.org/10.1016/B978-0-12-824320-6.00004-6

Considering the healthcare waste produced by different stakeholders, 15% of the total share of HCW is regarded as hazardous while 85% of the HCW is non-hazardous (WHO, 2018). What is more alarming in this regard is that most of the hazardous HCW are few times more hazardous than usual household waste (Padmanabhan and Barik, 2019). Thus, exposure to such waste leads to extremely negative outcomes comparing with the exposure to usual household waste. Though hospital waste is the most common form of HCW, it is not limited to only hospitals; instead, all the healthcare facility providers can be regarded as HCW producers. The amount and toxicity of HCW produced by healthcare industries is also comparatively higher than many other industries. Usually, 25 pounds of HCW is produced per day by each patients and as per the data of United States, to produce 1 pound of healthcare service, 32 pounds of waste is generated during manufacturing process (American Hospital Association, 2015). These estimations of waste generation and waste exposure vary significantly among countries as per their socioeconomic conditions (Minoglou et al., 2017). However, to deal with the HCW successfully, waste minimization should be prioritized over waste treatment, as the waste treatment process itself can be highly hazardous.

As HCW is highly hazardous, its prevention is a safer approach than its treatment. In doing so, the role of all the relevant industries is important. As many different types of HCW can be produced, their source-level reduction strategies should also be different. Being a highly fragmented industry, source-level reduction of HCW is extremely difficult, as small players are not often concerned enough to contribute in HCW source reduction (Minoglou et al., 2017). Besides, there is not much scope of supply reduction or reuse of healthcare equipment, as in many low income countries, the amount of healthcare services is already inadequate. As a result, introducing common waste source reduction practices such as reduction or reuse approaches might not be much effective in healthcare industries. Additionally, regulation on proper disposal of HCW is also quite ineffective because of having fragmented smaller players of healthcare industries (Padmanabhan and Barik, 2019). Considering all such difficulties and challenges of sustainable HCW minimization and source reduction, more sophisticated approaches should be introduced.

This chapter explores some of the commonly produced waste in the healthcare industries and discusses their corresponding healthcare source and biogeochemical hazards. Besides, this chapter also suggests some of the major source-level waste minimization approaches to prevent such HCW generation.

4.2 Commonly produced waste in the healthcare industries: source and biogeochemical hazards

Industrial waste can be classified in terms of different characteristics, that is, toxicity type, generation source, corresponding hazards, and so on. Similarly, healthcare industry produced waste materials can also be classified on several different basis. In this chapter, considering the huge diversity of healthcare industry produced waste, the characteristics and hazard potentials of HCW have been taken into account to classify waste produced in healthcare industry. This includes—infectious waste, pathological waste, sharp waste, chemical waste, pharmaceutical waste, cytotoxic waste, radioactive waste, genotoxic waste, general waste, and so on (WHO, 2018; Padmanabhan and Barik, 2019). These different classes of HCW are generated in specific fields of healthcare industry and each different sector of healthcare industry is responsible to generate at least more than one type or classes of HCW. The extent of HCW is so diverse that each of these classes also contain a huge varieties in terms of their hazard potentials, generation source, and certain properties. Such classification of healthcare industry generated waste helps to determine suitable storage and safeguarding, ensuring minimum exposure and hazards (Padmanabhan and Barik, 2019). However, discussions on HCW classes should not be limited only upon these classes, rather, this classification is expected to provide a systematic understanding in terms of varieties of HCW.

Table 4.1 classifies and discusses some of the commonly produced HCW, their healthcare sources, and corresponding hazards.

4.3 Source-level waste minimization initiatives by the healthcare industries

Since a lot of different types of wastes are produced by the small to large players of healthcare industry; mostly during utility generation and after using utility services, source level waste minimization initiatives of such healthcare industries should involve a variety of initiatives implemented right after the waste are generated in order to minimize waste at source-level. The greater variety, higher toxicity of waste classes, and numerous number of role players under this industry make waste minimization initiatives

Table 4.1 Major classes of commonly produced waste in healthcare industry (WHO, 2018; Padmanabhan and Barik, 2019; UNEP and WHO, 2002).

Class	Definition, facts, and properties	Healthcare sources	Corresponding hazards
1. Infectious waste	Infectious wastes are biohazardous or biomedical wastes which carry infection causing agents. Infectious waste carry biological or chemical agents that may cause disease in short term or in long term. Apart from healthcare services, any healthcare related activities might be responsible for infectious waste generation. Usually, 15%–25% of total produced HCW are infectious waste. Sometimes, some pathological or sharp waste can also be regarded as infectious waste because of having infection causing properties. A wide range of healthcare services, that is, diagnosis, treatment, prevention produce huge share of infectious waste.	Diagnostic samples of blood or body fluids, infectious agents released from autopsy, suction canisters, and animal body in laboratories, infected swabs, gowns, gauze, masks, bandages and medical equipment, microbiological samples, fetuses, body parts, tissues, infected or used containers, expired drugs, contaminated excreta and urine, used solutions, vaccine waste, etc.	• Infectious wastes are responsible for the occurrence of different infectious disease (Influenza, Hepatitis A/B, Tuberculosis, SARS, etc.) through transmission of microorganisms (viruses, bacteria, parasite, and fungi). • Environmental exposure of infectious waste destroys useful microbes, plants, and harm animals. • Even short-term exposure to infectious waste may cause occurrence of infectious disease. • Agriculture sector also gets affected by infectious waste causing damage of significant amount of agricultural product damage.

2. Pathological waste	Pathological wastes are a major type of infectious and biohazardous waste which are suspected or found to contain pathogens or can be easily transmit pathogens as a carrier are known as pathological waste. These wastes are always responsible to pose a risk of disease transmission by transmitting pathogens. Pathological wastes are highly hazardous since short time exposure or low density exposure can even increase risks of disease occurrence. Mostly surgical procedures involve significant production of pathological waste.	Human body fluids, blood, organs and tissues, body parts, animal carcasses, suction canisters, gowns, cotton/gloves contaminated with blood or body fluids, gowns, gauze, masks, tissues, fetuses, body parts, microbiological culture or broths, infected containers, expired drugs, contaminated excreta and urine, used solutions, vaccine waste, etc.	• Pathological wastes are responsible for pathogen transmission-mediated diseases. • Pathological waste may also lead to infections sometimes based on the types of the pathogens that are being transmitted. • Pathological wastes are responsible for contagious disease transmission. • Contaminates equipment enhancing the risks of infecting others.
3. Sharp waste	Sharpe wastes include at least one to few sharp edges. HCW utilities often have sharp edges, regarded as sharp HCW comprising 1% of total generated HCW. Injections are the most common sharp waste category and annually 16 billion injections. They might be responsible for physical hazards and sometimes work as a carrier of infectious or pathological waste transmission. Especially infections of HIV, hepatitis B/C sharp unsafe injection usage. Apart from this, needles, equipment, and other sharp waste are produced in large number every day by diagnosis centers, clinics and by other healthcare service providers.	Blades, needles, syringes, disposable scalpels, broken glass, pipettes, cover slips, slides, petri dishes, saws, intravenous set, broken ampoules. and vials, etc.	• Sharp waste transmits some specific disease-causing agents, that is, tetanus, etc. • Sharp waste is mostly responsible for physical hazard occurrence such as bleeding, scratching, etc. • Usage of contaminated sharp waste (i.e., syringes or needles) also cause HIV/Hepatitis transmission. • Improperly managed sharp waste often cause occurrence of accidents.

(Continued)

Table 4.1 Major classes of commonly produced waste in healthcare industry (WHO, 2018; Padmanabhan and Barik, 2019; UNEP and WHO, 2002). (*Cont.*)

Class	Definition, facts, and properties	Healthcare sources	Corresponding hazards
4. Chemical waste	Unlike other industries, healthcare industry also includes a huge share of chemical waste. Chemical wastes are those wastes that contain environmentally harmful and health-wise hazardous elements and chemical compounds. Chemical waste produced in one sector of healthcare industry might not be similar to the chemical waste produced in another sector of healthcare industry. Altogether, 2%–3% of total generated HCW are chemical waste. It causes emergence of huge amount of chemical waste making it more difficult to adopt suitable minimization strategies.	Chemicals used in health research and diagnosis laboratories, batteries, disinfectants, sterilizers, medical devices (MRI machine, CT scanner, etc.), and equipments (thermometer, metal rings, etc.) rich in heavy metals or alloys, chemotherapeutic agents, spent disinfectants, fixatives, formalin, solvents such as xylene, toluene, methanol, etc., broken thermometers, etc.	• Depending on the type of chemical waste, it can be often responsible for organ dysfunction, skin irritation, and others. • Cause significant amount of soil, air, and water pollution. • Initiates environmental degradation and biodiversity loss. • Improper management of healthcare chemical waste may lead to cancer, tumors, or other deadly health conditions.
5. Pharma-ceutical waste	Pharmaceutical waste denotes all the drug-production, distribution and disposal related waste. Pharmaceutical industry itself is a huge industry which can produce many different wastes including infectious, pathological, cytotoxic, and radioactive waste. However, in this subsection, mainly pharmaceutical drugs have been categorized as key pharmaceutical waste, as other types of pharmaceutical industry produced waste overlap with some other classes of HCW. Around 2% of the total generated HCW are pharmaceutical waste.	Vaccines, unused and expired drugs, etc.	• Degrades air and water quality • Kills useful microbes, animals and also degrades human health condition. • Hampers biogeochemical parameters. • Such degradation of valuable resources leads to significant monetary loss.

6. Cytotoxic and genotoxic waste	Cytotoxic and genotoxic waste often overlap in terms of healthcare industry. Cytotoxic waste contains residues that are toxic at a cellular level. And genotoxic wastes usually have mutagenic, carcinogenic, or teratogenic properties. In other words, genotoxic wastes are toxic at a genetic level. Such wastes are highly hazardous and exposure of the slightest amount can be extremely harmful. Less than 1% of totally generated HCW are cytotoxic or genotoxic. Storage and disposal of such waste should be maintained under very strict regulations.	Anti-cancerous drugs, cytotoxic and genotoxic drugs, broken thermometers, spent disinfectants, anesthetic gases, etc.	• Initiate cellular deformation, tumor formation, and occurrence of cancer. • DNA damage. • Occurrence of genetic disorders. • Cytotoxic and genotoxic properties are extremely hazardous; both for human and animals for which even short time exposure to little amount may lead to death.
7. Radioactive waste	Radioactive wastes are those form of waste, which contain radioactive substances or radionucleotides. Such radioactive waste are below 1% of the total generated HCW. However, such smaller percentage doesn't represent its corresponding hazard potentials since harms caused by radioactive waste are 100–1000 times more dangerous than other waste categories. Strict regulation of such waste is a must to protect individuals.	Radionuclide contaminated medical instruments, radiotherapeutic materials, radioactive diagnostic materials, chemotherapeutic agents, radioactive elements (usually used for imaging or therapeutics), etc.	• May cause cellular or genetic deformation. • Radioactive radiation leads to long term damage of organs and physiological processes. • May lead to deaths of human and animals.

(Continued)

Table 4.1 Major classes of commonly produced waste in healthcare industry (WHO, 2018; Padmanabhan and Barik, 2019; UNEP and WHO, 2002). (*Cont.*)

Class	Definition, facts, and properties	Healthcare sources	Corresponding hazards
8. General waste	General wastes are those category of waste which do not have any infectious, cytotoxic, radioactive, or any other properties mentioned in the above classes. Though many intellectuals often regard this category of HCW as non-hazardous, in a critical lens, none of the waste categories are non-hazardous. Instead, in case of such waste class, the corresponding hazards or impacts are quite indirect and often their environmental impacts are predominant over their direct health impacts. However, in most of the cases, such category of waste do not cause extreme health situation that other HCW classes (radioactive, cytotoxic or infection waste) may cause. Some of such wastes are flammable, corrosive, and explosive. Over 75% of HCW is general waste making this category as the largest HCW class.	Packaging, paper, food scrape, container, bottles, diaper, bags, uncontaminated masks, hats, gloves, shoes and gowns, cardboard, wood, metal, etc.	• Increases environmental burden of waste. • May cause indirect health impact. • Improper management or inadequate initiatives lead to extremely negative impacts in the long run.

Figure 4.1 ***Overview on the source-level healthcare waste minimization strategies.***

even more difficult. Fig. 4.1 illustrates some of the major source-level HCW minimization initiatives.

4.3.1 Waste segregation

Waste segregation is a very basic waste management mechanism that denotes separating waste of different classes, as per their characteristics, corresponding hazardous properties, and recycling potentials. This might not commonly seem a very effective waste minimization initiative, however, it is found to be significantly effective in regulating HCW exposure in our society. Some of the common purposes of waste segregation include safe disposal, proper recycling, and most importantly, minimizing waste. It is found that only basic segregation of HCW can reduce 2%–25% of HCW generation (HealthcareHarm, 2020). And it reduces 2%–5% of HCW-borne hazards (UNEP and WHO, 2002). This denotes the huge potentials of waste segregation mechanism as a waste minimization technique. In addition to it, separating different hazardous waste categorically soon after their generation reduces disposal cost and increases raw materials for recycling (UNEP and WHO, 2002). Besides, since HCW segregation takes place at the source-level, there is almost no chance to transmit waste-borne negative impacts, if they are disposed securely and immediately after generation. This is how HCW segregation also contributes in other waste minimization approaches confining waste at a limited area. However, despite being a very simple system to implement, many healthcare industry-related stakeholders are not concerned enough to assure proper waste segregation due to their ignorance, indifference, and reluctance. Such situation is common among the healthcare industries of developing countries whereas the developed countries are found to successfully implement proper HCW segregation at source-level leading to a sharp reduction to HCW-borne hazards.

Apart from the country-wide disparities in terms of proper HCW segregation, there are several other challenges that are specific to healthcare waste segregation. First, as waste-related knowledge of healthcare staffs varies significantly, there is a large share of healthcare staffs which are not well-informed regarding their role and responsibilities in terms of waste segregation. For instance, in a leading tertiary healthcare service providing hospitals of Nigeria, only one-tenth of the health staff were provided training on their role in waste minimization (Abah and Ohimain, 2011). As a result, in places where such trainings are similarly inadequate, most of the healthcare staffs often ignorant to install HCW segregation facilities properly and sometimes even if waste segregation facilities are available, many of them are reluctant to use these facilities properly. Second, sometimes even if the wastes are segregated properly, the future of segregated waste is quite unpredictable. For instance, in Bangladesh, though many of the hospitals are practicing well-facilitated waste segregation, these segregated wastes do not undergo proper recycling approaches and in many cases, these segregated waste are disposed mixed with other waste without proper safety measurements (Nuralam et al., 2017). Thus, the segregation process does not bring any positive outcome, instead increase its exposure to the environment. Besides, the smaller and commercial players of healthcare industries are not often motivated enough to practice waste segregation, as their major purpose is money-making out of their business for which they are hardly concerned about HCW's corresponding hazards. This situation gets even worse when these players are not encouraged to contribute in HCW segregation through policy adoption. Thus, at different contexts, HCW segregation initiatives face multiple challenges.

However, to overcome different challenges associated with HCW segregation, primarily it is necessary to launch suitable waste segregation system by the healthcare industrial role players. In this regard, the color coding and the labeling systems are widely implemented (UNEP and WHO, 2002). The color coding system is a very simple system where waste collecting containers used in healthcare industries are colored with black, brown, or yellow color. Each of the colors has different meanings. For instance, less hazardous and general pharmaceutical wastes are disposed in the black containers; infectious, highly toxic and radioactive wastes are disposed in the yellow container; chemical and metallic wastes are disposed in the brown containers. In this regard, the highly infectious waste are first stored in polythene bags and then disposed in the yellow bins to assure more safety. This is how color coding system differentiates waste categories for waste segregation.

On the other hand, labeling system is also another popular waste segregation approach. In this system, an internationally used sign is assigned for every waste category and corresponding waste containers are marked by these different signs so that the stakeholders can easily dispose respective type of waste in the corresponding containers. However, all these systems will be of no use, unless the healthcare staffs are trained how to use these systems properly. This is why necessary training and knowledge dissemination should be ensured. Aside from these, the healthcare industries should be encouraged to practice waste segregation by government interventions and suitable policy adoption.

4.3.2 Product substitution, process modification, and technological advancement

Like other industries, waste minimization and source reduction initiatives at the healthcare industries should also incorporate significant amount of product substitution and process modification to ensure minimization of HCW at source level. Such source reduction mechanism is quite tough to be implemented in this particular industrial sector, since there is almost no alternative or safer option to several infectious, radioactive or genotoxic waste generation. Instead, in case of such waste, immediate postgeneration waste minimization can be regarded as suitable waste source reduction mechanism. However, there are yet several scopes for product substitution and process modification in this particular context. Especially, in case of developing countries, many of their unsustainable and toxic waste producing healthcare industrial practices can be substituted by safer alternatives. In such countries, less quality products are often used to provide healthcare support that produce comparatively more hazardous and more amount of HCW (Abah and Ohimain, 2011). On the other hand, the product quality and processes involved with healthcare industries are regulated by suitable policy instruments and regular evaluation in the developed countries. Therefore, product substitution and process modification initiatives are more urgent in the developing countries incorporating modern technologies, tools, and approaches.

While substituting healthcare products or modifying healthcare industrial with an aim to source-reduction of HCW, some of the considerable features should be its sustainability, effectivity, and corresponding generation cost. Besides, reducing, reusing, and recycling features can also be incorporated to determine a suitable substituent of the products. In this regard, the example of widely used disposable surgical attire can be a good example.

Disposable surgical attires (gloves, gowns, etc.) make a good share of total HCW and in many cases, they are preferred over reusable surgical attires. However, if substitution of such attires by reusable surgical attires is ensured, a large share of highly hazardous HCW generation will be reduced. However, reusable attires should be cleaned and autoclaved properly before repetitive usage; therefore, installing autoclave machine is necessary before introducing reusable surgical attires. It not only reduces waste production, at the same time, it reduces expenditure spent for surgical attires due to long-term durability (McHugh et al., 2014). At the same time, it is often found that in low-income countries, due to inadequate resource allocation in hospitals, healthcare staff are often compelled to use one-time surgical attires for multiple times (Cawich et al., 2013). In this regard, if one-time expenditure on reusable surgical attires along with suitable autoclaving facilities is ensured, source-level waste reduction of such highly toxic HCW will be possible. Likewise, whenever there are other scopes to shift from disposable to reusable goods.

Substitution of products can also be encouraged through drug substitution at different level. This may occur in various ways. For example, if introduction of HCW minimization-related monitoring and evaluation services reveal that any pharmaceutical industry is failing to adopt enough initiatives to minimize their generated waste, their produced drugs might not be allowed in the market at a large scale. Though this may lead to a negative impression for many patients who often require drugs of that specific company, but under this regulation, a substituted drug produced by another company might fulfill the consumer's demands (Kjoenniksen et al., 2006). This strategy will encourage the pharmaceutical industries to adopt suitable waste minimization and source reduction strategies. The government can subsidize or provide extra trading facilities for the companies that will introduce less waste generating processes. Besides, drug substitution may occur at an industrial level. If any drug is found to be responsible for a higher amount of toxic waste generation, this drug can be substituted by a less waste producing drug. Though, this might not be possible in all the cases due to limitations of innovative process designing. But industry-level drug production process needs to be ungraded at a regular basis so that efficiency of the generation process and efficacy of the drugs are ensured (Gaynor et al., 2007). While permitting a drug in the market, its corresponding generation processes and impacts should be assessed to ensure sustainability of the process.

Aside from drug substitution, processes involved with different treatment processes such as therapies can also be substituted by safer options with an aim to minimize corresponding toxic waste generation. For example, chemotherapy and radiotherapy produces a significant share of highly toxic chemical and radioactive waste. Though in many of the clinical cases, such therapies do not have any suitable alternatives, with more innovative efforts, their extent and exposure can be limited. Since such therapies often fail to detect a precise area to provide the therapies and damage other non-infected cells, researchers are trying to discover innovative approaches to detect infected cells more precisely (Hasegawa et al., 2003). At the same time, reducing the amount of such highly toxic agents is often a matter of concern for many researchers. As a result of such initiatives, it is hoped that less amount of chemical or radioactive agents will minimize significant share of corresponding highly hazardous waste. Besides, if possible, other disease diagnosis and treatment processes should also be modified to assure minimum amount of waste generation. Disease diagnosis or pharmaceutical processes using highly toxic carcinogenic or genotoxic agents should be substituted comparatively safer options. Many of such initiatives are being noticed even in the low income countries where subsequent amount of resource limitations are noticed. Especially in the pharmaceutical industries there are adequate numbers of options to replace many hazardous chemicals by suitable non-hazardous chemicals while producing certain drugs (Subrata et al., 2013). Disease diagnosis can also be less waste producing through suitable chemical substitutions; for instance in modern days gold nanoparticles are often used to detect genetic disorders such as Alzheimer's disease (Han et al., 2019). Such substitutions are more environmental friendly and safer for human health due to the wide range of positive characteristics of nanoparticles.

Process modification is an equally important source-level waste minimization approach along with product substitution for different industries, since inefficient industrial processes are often responsible for large amount of industrial waste generation. Likewise, healthcare industry related processes should be modified and upgraded constantly to determine the most efficient and the most sustainable process so that a large share of HCW generation can be avoided. Constant innovative inclusions in the healthcare industrial processes facilitate rapid growth and in most of the cases, such inclusions promote sustainability, cost-effectiveness, and other co-benefits (Thakur et al., 2012). Among many of the potential process modification options to be adopted for sustainable source-level waste minimization, primarily, the most efficient manufacturing/diagnosis/preventative initiatives should be adopted. Hindiyeh et al. (2018) argued that comparison of environmental impact assessment for every pharmaceutical processes should be conducted, since slight substitution of chemicals may largely cause variation in terms of their waste generation and emission potentials. Therefore, different options-related with industrial processes should be compared to determine the most suitable process and thus the corresponding process should be modified.

Apart from comparison based large scale industrial process modification, different healthcare related strategical options involved with disease diagnosis and treatment should also be assured. Disease diagnosis and treatment strategies are one of the most responsible sectors of the healthcare industry causing generation of the most hazardous waste. However, there is no "one best strategy" related with healthcare industry (Al-Abri, 2007). Instead, the best healthcare industrial strategy should be explored as per suitable context of the corresponding industries. However, since many of the clinical processes involve generation of significant amount of highly toxic waste, some source-level initiatives should be adopted in this area to minimize the amount and toxicity of such waste. For example, the community level primary level healthcare service providers of the tropical developing countries often lack good amount of necessary facilities and in most of the cases, they lack suitable incineration and sterilization facilities causing generation of a large amount of infectious or pathological waste. As a result, along with these waste, reusable sharp waste cannot also be sterilized properly in many cases (WHO, 2005). Inclusion of sterilization facilities or supply of enough amount of disinfectants are hoped to improve the situation.

Process modification initiatives can be as simple as sterilization facility inclusion to modifying related machineries. For instance, if two companies launch a similar type of radioactive material using machinery of different efficiency in terms of radioactive material usage, the less efficient machine should be substituted by the efficient one. This is reduce generation of radioactive waste at a source level and it will also minimize the recurrent cost of radioactive material usage. Another potential process modification initiative aimed at waste source reduction might be installing resource recovery technology before waste disposal. This is more applicable for large scale healthcare industries where subsequent amount of valuable resources are disposed due to insufficiency or ignorance in terms of resource recovery facilities (Velagaleti and Burns, 2007). Such resource recovery initiatives will mostly reduce chemical waste generated from healthcare industries. Though such recovery process installation will incorporate good amount of capital cost, however, its long-term outcomes will compensate the capital cost ensuring large share of cost reduction in terms of resource purchase and environmental sustainability associated with the process. Yang et al. (2009) argue that such healthcare process modification initiatives may cause even worse impacts, unless suitable technological verifications, environmental monitoring or supervision, training module designing, etc. are ensured beforehand. Especially, unless the healthcare staffs are trained on-hand in terms of utilizing the modified processes, this whole operation will be of no use (Tudor et al., 2005). Therefore, even if process modification initiatives are introduced, adequate amount of preparation and evaluation are needed at the same time.

In addition to different product substitution and process modification initiatives, in different healthcare service provider industries, dependency on modern technologies should also be increased, since in most of the cases, older technologies are less efficient and more waste producing processes. This is a part of process modification that modifies healthcare-related industrial processes with an aim to minimize waste generation at source level. In this regard, inclusion of components of modern information technology in healthcare sector is a very popular concept around the globe. Applications of information technologies in multiple healthcare-related industrial sectors offer several benefits that include cost reduction, waste minimization, higher accuracy, increased efficiency, easier maintenance, sophisticated supply chain management, and others (Bates, 2000; Helms et al., 2008; Bhattacherjee and Hikmet, 2007). All these benefits out of information technological interventions in healthcare sectors benefit all the stakeholders in multiple ways. For instance, mobile app based consultancy on expert physician offers utilization of significantly less amount of resources and this is an indirect source level waste minimization mechanism derived from process modification initiatives (Imteaj and Hossain, 2016). Besides, from the healthcare service providers' point of view, increased dependency on IT facilities makes their activities more organized and less time consuming assuring consumption of less resources and less production of waste materials. Though there is a mixed reaction of the healthcare service providers regarding their stand in terms of expanding dependency on IT, it is yet regarded as one of the most sophisticated and safest strategies to overcome many of the industrial limitations (Bhattacherjee and Hikmet, 2007; Poon et al., 2006).

For source level waste minimization, technological dependency should not only be limited to expansion of IT facilities. Instead, other technological inclusions should also be assured at the same time for sharp reduction in waste generation. In this regard, RFID tracking system can be a relevant example to be implemented in the healthcare industries. Here, RFID denotes Radio Frequency Identification and its incorporation in healthcare industry will offer a variety of facilities assuring better monitoring, evaluation, necessity detection and necessity detection (Lee and Shim, 2007). As discussed in the previous chapters, RFID tracking system is often implemented for more efficient waste minimization, such system can also be implemented in healthcare industrial waste minimization. This system is comparatively more applicable for instantaneous response against waste generation. This is an indirect source-level waste minimization strategy, as it helps to assure waste prevention immediately after generation; especially for the waste classes whose generation cannot be avoided. At the same time, better segregation mechanism can also be designed under RFID tracking system. In addition to it, mobile-based applications help to design better personalized healthcare at the shortest possible time using limited resources (Kalem and Turhan, 2015). Apps and software also ensure data analysis, disease diagnosis, diet designing, and other processes easier ensuring less instrumental or technical involvement that is another technology-based source-level waste minimization initiative. At the same time, introducing more green analytical chemistry-related technology in health sectors minimizes chemical waste generation significantly (Gałuszka et al., 2013). Thus, technological innovations often assure source-level waste minimization providing cleaner and less waste producing alternative options.

4.3.3 Reducing, reusing, and recycling

The healthcare industry is a highly toxic waste generating industry compared to the toxicity of waste generated in the other industries. Therefore, introducing 3R approaches in case of healthcare industries is tougher than other industries, since many of its generated waste, such as infectious, pathological, carcinogenic, or genotoxic waste, are not even recyclable. However, yet, there are some scopes to introduce source-level 3R approach in the healthcare industries, especially for the waste class of general waste, sharp waste, and others. However, there are evidence of successful recycling of sharp waste with an aim to hazard minimization and resource recovery. For example, with an aim to source-level sharp waste recycling, WHO (2005) launched pilot project to assess safety, viability, and cost-efficiency of recyclable auto-disable syringes during immunization programs which was quite successful. However, such success factors were not only dependent on a shift from disposable to recyclable syringes, instead, chemical disinfection, manual needle separation, etc. should also be maintained strictly during such recycling initiatives. Besides, since such auto-disable syringes are not that available around the globe, healthcare industries may also take initiatives to isolate necessary elements from the used sharp items for future manufacturing (Chen and Chen, 2003). However, in this process of recycling, sophisticated segregation, and proper contextualization of initiatives are needed; otherwise there are chances of transmission of infectious and pathological agents.

Apart from the recycling of the sharp waste, for sustainable recycling of general waste, a variety of recycling approaches can be adopted. As discussed in Table 4.1, due to the direct involvement of a large number of stakeholders in this industry, over three-fourth of the produced HCW are general waste in the healthcare industries most of which can be recycled, composted or even reused often. Such general waste do not carry much infectious or pathological agents in most of the cases for which recycling such waste is comparatively safer than recycling other classes of HCW. First, uncontaminated HCW are often categorized under this class. For instance, unused torn gloves, gowns, or other safety materials are often disposed which can be fixed as needed and can be reused. Besides, dysfunctional machineries are often disposed as general HCW. If there is no scope to fix the machineries, its materials can be isolated and recycled to produce other valuable goods. Such machineries often contain a good amount of metallic and organic pollutants for which if there is a lack of initiatives to recover useful metallic or organic resources from such machineries, these pollutants will be exposed to the environment and thus they will bring about severe short and long term consequences. Most importantly, the functional parts of the dysfunctional machineries also have market value so the healthcare industries should introduce departments that will be solely responsible to determine innovative ways for minimizing such waste generation.

Composting can also be regarded as a suitable recycling initiative for the healthcare industries. As discussed in the previous chapters, composting is the process of decomposing organic matters under suitable conditions so that it produces necessary soil conditioners and nutrients. Composting can be a suitable source-level waste minimization approach for minimizing organic waste of the healthcare industries; especially for tropical regions or where there are scopes to provide suitable composting conditions. For HCW minimization, composting can be a good option, since it helps to destroy harmful agents and at the same time, recovers useful resources (Tudor, 2007). Composting can be useful to destroy infectious, pathological, and some general waste. Especially in case of promoting the concept "Green

Hospital," assuring compositing facility is a must. Here green hospital denotes such a term that prioritizes the welfare of both the human health and the environment (Azmal et al., 2014). Composting facility can be introduced nearby or at a need-based location. Especially in case of HCW composting, source-level composting should be prioritized since such waste are highly hazardous and transportation of such waste for composting would cause transmission of harmful agents. For HCW composting, organic matter-rich waste are decomposed in an oxygen-rich environment and in this process some of the commonly treated waste might be hospital food waste, paper, boards, containers, old infected non-metallic containers, wooden furniture, and others. If the healthcare industries are unable to do so, external agencies can be assigned to conduct such initiatives.

To successfully implement different reusing and recycling initiatives, reducing, the basic step of 3R approach implementation, should be prioritized. It is because reduction on resource usage broadens up scopes to assure successful implementation of reusing and recycling initiatives. Usually, many of the countries are struggling very badly to set up sustainable resource allocation for the healthcare industries. Though the potential reasons behind such uneven and unsustainable resource distribution vary from context to context, it is found that in many cases necessary healthcare resources are inadequate and often unnecessary healthcare resources are distributed (Chaerul et al., 2008). Especially the developing countries are the worst victims of such poor healthcare resources distribution. As a result of such poor management in terms of resource distribution, a huge share of healthcare instruments becomes dysfunctional and converted into waste in later terms. For example, in Bangladesh, improper resource allocation is quite common in the health sector due to lack of need-based resource distribution (Rahman, 2019). If need-based resource allocation is ensured, a large share of waste generated from unused resources will be minimized. This can be regarded as a source-level application of 3R approach in the healthcare industry. Besides, waste reduction strategy should also make sure not to utilize unnecessary resources so that infected waste generation can be prevented. Especially, strict regulation should be maintained to confine infections in assigned regions; for instance, surgical wards, chemical laboratories, and other departments of healthcare industries should be limited only for respective individuals and they should also maintain proper hygiene and disinfection strategies while leaving such infected areas. This is a waste reduction strategy since it reduces unprecedented transmission of hazardous waste.

Despite there are limited scopes to implement 3R approach in the healthcare industries, implementation of 3R approach in this particular context, being one of the most diverse industrial setting, should contextualize different 3R approach-based initiatives as per their co-current context, available facilities and sustainability impact assessment. Especially, the regulative authorities of the healthcare industries need to play an important role for successful implementation of different initiatives under 3R approach and the government bodies should take part in this process actively to ensure better industrial contribution in this sector.

4.3.4 Thermal treatment

Since the healthcare industries are responsible for the generation of a variety of infectious and pathological waste, source-level thermal treatment of such waste can be a potential short term solution to assure source-level minimization of HCW. It is because thermal treatment offers a variety of methodologies to destroy infectious agents, pathogens, chemical toxins, and others. A variety of different thermal treatment facilities might be applicable for healthcare industries that include autoclave, sterilization, incineration, and others. As per the industrial contexts and available facilities of the healthcare industries, forms of suitable thermal treatment might vary. All these different thermal treatment approaches offer different facilities and at the same time also incorporate several unavoidable limitations. As per the healthcare industrial necessity, thermal treatment initiatives help to largely minimize both the extent of toxicity and the amount of the HCW. Though it is not a pregeneration waste minimization strategy, immediate thermal treatment after generation offers nearly source-level waste minimization. Unfortunately, thermal treatment requires a lot of energy consumption and is not much environmentally sustainable. However, sophisticated modifications using efficient technology may increase sustainability of the thermal treatment processes.

Among many different thermal treatment processes commonly implemented in the waste minimization sector, autoclave and incineration are the most popular ones. Potential reasons behind the popularity of such approaches include low implementation cost, process simplicity, no technical complexities, and others. Here, autoclave is the instrument used for thermal treatment process that facilitates industrial or scientific processes at an elevated temperature. Using autoclave is a suitable strategy for thermal treatment of infected healthcare-related equipment. Autoclave machine can kill infectious agents and thus, can sterilize or disinfect clinical or laboratory equipment before usage (Ghasemi and Yusuff, 2016). Besides, autoclaving healthcare waste before disposal; especially the highly infectious waste, helps to minimize the corresponding hazards and toxicity of the HCW. Even for some inorganic waste removal mechanism from HCW, autoclave can be regarded as a suitable option (Pinho et al., 2015). However, some of the limitations of autoclave are energy inefficiency, inability to be implemented at a large scale, insufficient to treat all harmful components from waste, and so on. Yet, for small scale and for highly infectious waste minimization, autoclave can be considered as a good option.

Incineration, being another popular thermal treatment strategy for healthcare waste minimization, can be considered as another realistic waste minimization option for different contexts. Incineration is the process where waste materials are burnt at high temperature in presence of oxygen. It offers a broader range of facilities over other waste minimization strategies that include no technical complexity, significantly low implementation cost, no necessity of complex set up, and others. Some of the negative aspects of this easily implementable thermal approach are its significant carbon emission potentials, environmentally unsustainable properties, and others. However, waste minimization initiatives in the healthcare industries might be significantly benefitted by incineration approach, especially in the low-income communities where installation of expensive waste minimization strategies is not possible. For instance, since installation of suitable clinical waste treatment facilities was not possible for the Kutupalong Rohingya refugee camps as an urge to the emergency response, the clinical waste were incinerated. Despite its environmental drawbacks, it can be regarded as a simple means of highly hazardous waste minimization (de Vries et al., 2017). Besides, sometimes healthcare waste are highly infected for which transporting them in suitable treatment sites becomes very risky for which incinerating such waste right after generation is often regarded as a suitable solution. Therefore, despite not being a core source-level waste minimization strategy, such different thermal treatment might benefit source-level HCW minimization, if immediate response right after generation is assured.

4.3.5 Biochemical treatment

As discussed in the previous chapters, different biochemical mechanisms can be adopted for source-level or immediate postproduction level waste minimization. Such biochemical processes involve incorporation of different physical, chemical, and analytical strategies. Implementing biochemical treatment mechanisms for HCW is quite difficult for the healthcare industries due to the involvement of a large number of stakeholders and presence of numerous small role players. Suitable biochemical processes to be implemented in different sectors of the healthcare industries should vary since the type of treatment needed for the pharmaceutical industries or the diagnostic centers are quite different from hospitals. For example, pharmaceutical industries may introduce multiple filtration or electrolysis mechanisms with an aim to recover certain resources whereas hospital-based waste minimization should introduce chemical disinfection or sterilization of waste instead of prioritizing resource recovery initiatives (Rosman et al., 2018; Ozder et al., 2013). However, biochemical treatment processes offer so many varieties of options that suitable biochemical process adoption may also help to minimize highly hazardous radioactive, genotoxic, and other waste (Osmanlioglu, 2006; Sharma et al., 2013). Sometimes, even if biochemical processes fail to fully treat such waste, they can significantly minimize the amount and hazards of such waste. In this regard, availability of suitable technologies should also be crosschecked at that particular location to assure uninterrupted future treatment facilities (Singh et al., 2016). Thus, as per the industrial necessity and availability of technologies, suitable biochemical waste minimization strategies should be introduced.

However, unfortunately implementation of such biochemical mechanism-based waste minimization strategies involves several difficulties along with their huge potentials of source-level waste minimization. Healthcare industry is a very fast-growing industry which deals with a very sensitive issue of health situation improvisation. Since this industry involves mostly healthcare-specialists, businessmen, government agencies, and customers, this industry hardly prioritize recruiting experts with good experience of waste minimization related biochemical approaches (Gerwig, 2005). Such reluctance and ignorance of the industrial authorities itself don't often allow suitable biochemical process installation at the industrial level. If suitable biochemical treatment initiatives are undertaken at an industrial level, it will help to minimize postgeneration waste minimization as well as assure process modification to prevent waste generation. Therefore, it is necessary to adopt biochemical waste minimization strategies that will help to minimize the HCW on-site ensuring minimum exposure.

4.3.6 Laws, regulations, and policies intervention

Source-level waste minimization for every industry needs to be regulated under laws, regulations, and policies. Such regulations minimize waste generation promoting other waste minimization strategies. Unless promotion of waste minimization strategies are assured through policy interventions, such technical waste minimization initiatives will be of no use, since the role players of the healthcare industries are not bound to adopt the waste minimization. Besides, since these industries are the major contributor of HCW generation, they themselves should take initiatives to minimize waste under different regulative strategies. However, policies for healthcare industries denote designing suitable strategies that are bound to be abided by the healthcare industries. Expert policymakers on that relevant fields design policies providing clear guidelines for different healthcare industrial activities and operations with an aim to assure safety, sustainability, regulations, and equity (Lluch and Kanavos, 2010). Along with these, healthcare policies are also hoped to assure better market situation, positive impacts on globalization, and other materialistic welfare in both short and long term (Tritter et al., 2009). Especially, healthcare policies and regulations related with waste minimization sectors try to impose industrial responsibilities in minimizing waste so that long-term sustainability and safety are assured. Thus, adoption of several healthcare policies focused on specific area such as waste source reduction, are designed to achieve a wide range of beneficial outcome.

First, subsidization can be a suitable measure to encourage healthcare industries for installing source-level waste minimization initiatives. Government can take initiatives to subsidize the industries to install waste minimization initiatives. Though it is true that healthcare industries are one of the highest subsidization receiving industries for many countries, however, only a very small share of this subsidization are dedicated toward waste minimization initiatives. Especially in case of the developing countries, only a very less amount of subsidization are offered by the government, therefore, the healthcare role players put the waste minimization initiatives at the very bottom of the priority list (Moreira and Günther, 2013). Even in many cases, instead of subsidizing the healthcare industries, other agencies are subsidized so that they can collect HCW and adopt suitable initiatives. Though it is a good initiative, there are chances of higher amount of exposure in this waste collection, transportation, and treatment process. Especially, when the municipalities of the developing countries are assigned to take the role of HCW minimization, the municipality authorities often struggle to assure suitable treatment mechanism for the highly toxic healthcare waste (Thanh and Matsui, 2011). This is why instead of subsidizing waste treatment initiatives, if the healthcare industries themselves are subsidized to adopt source-level waste minimization initiatives, there are less chances of exposure and more possibilities to assure immediate postproduction treatment of the highly hazardous waste.

Apart from subsidization, the healthcare industries that take responsibilities to reduce subsequent share of their produced waste minimization should be incentivized under different schemes of standardization. Incentives are one of the major factors that encourage the stakeholders to adopt initiatives. Especially in case of healthcare industries, if the private sectors are incentivized, there will be more industrial liabilities in reducing the waste. Yang et al. (2009) argue that providing incentives to the industries promote source-level waste minimization initiatives more readily. At the same time, standardization of the waste quality is also important. In this regard, a minimum standard should be maintained that all the healthcare industries are bound to maintain. Healthcare operations should also avoid using the highly hazardous substances, if there are options for safer alternatives (De Regge et al., 2019). This should be regulated under government's regulations and government must impose strict regulations against the usage of the highly hazardous chemicals. This will also open up multiple cost savings opportunities and improved safety maintenance options for the healthcare industries in the long run. For instance, drug and therapeutic waste minimization in several European hospitals' oncology departments was found to be a suitable measurement to reduce expenditure and severity associated with the cancer patients (Fasola et al., 2014). Even, changes in the healthcare service providers' environmental legislations and immediate response in reducing healthcare waste, largely reduced cost and improved service quality in different healthcare industries of England (Tudor et al., 2008). Thus, standardization of a minimum waste quality before disposal, substituting highly hazardous chemicals by safer alternatives, and designing incentives under strict government regulations—all these help to assure source-level waste minimization.

In addition, government should arrange or encourage the healthcare industries to assure suitable staff training in this specific area of source-level waste minimization so that the staff gets the opportunities to understand their respective roles to minimize waste at source-level. Planned staff training will make the whole HCW minimization related operations easier to be implemented. However, since HCW minimization policies introduced by the government often prioritize training and educating the healthcare staffs, their performances, practices, and perceptions improved largely. As a result, the proactive role of the trained staff can assure more efficient waste minimization (Gerwig, 2005). Staff training in the developing countries is even more needed, since many of the healthcare industries of the developing countries cannot afford expensive waste minimization initiatives that can be substituted by sincere staff efforts. In the developing countries like Nepal, hospitals' interventions through staff training and planned source-level minimization of daily produced sharp, chemical, and infectious waste significantly reduced occurrence of smaller to larger accidents and ensured an overall improvement in the hospitals' environment (Sapkota et al., 2014). This will be even more impactful when the government intervenes in this regard by designing suitable staff training opportunities. The government should also undertake some of the responsibilities to train up as much healthcare staff as possible so that quality training can be assured. In this way, the staff themselves will contribute in proving their organizational policies. At the same time, more enforcement of laws and regulations are needed for quality assurance. Thus, government regulations and interventions are hoped to benefit the stakeholders to promote suitable source-level waste minimization initiatives.

Case Study 5: Sustainable Medical Waste Minimization Plan by "Reduction in Motion"

Since medical waste is one of the most toxic waste generating industries, sustainable management of such waste is truly important to initiative source-level waste minimization. However, the initiatives should not be confined only toward the government's initiatives, since the amount of medical waste produced everyday is quite large. Therefore, sincere role of the private sector should expand in this sector. Among many of the notable private sectors, role of different firms can be illustrated for sustainable medical waste minimization strategy designing. Reduction in Motion can be regarded as a suitable example in this context. This is a sustainability consulting agency, focused in the healthcare sector. They have been designing suitable medical waste minimization strategies and plans for over 15 years. In this process of context-wise strategy designing, at the initial level they analyze the field-based context and then gradually incorporate healthcare waste minimization initiatives for their customers. Cost savings is one of their major priorities since they argue that usually conventional minimization of HCW requires 17 times higher costs than other industries (Reduction in Motion (2020)). Therefore, they try to offer the most suited and most affordable strategic plan to minimize medical waste.

The offered services of this firm include primary diagnosis of the situation, categorizing or prioritizing the necessities, and finally introducing treatment phases. Additionally, Reduction in Motion promotes sustainability of the waste minimization initiatives by offering need-based training programs and workshops. Another unique feature of this firm is to follow-up or impact assessment (Reduction in Motion (2020)). This is also an important step since unless proper follow-up is conducted, determining the effectivity and designing suitable modifications will not be possible. Thus, this private sector initiative is also contributing for source-level HCW minimization through various approaches. Considering the future prospects and urgency of such initiatives undertaken by the private sectors, this can also be considered as an emerging business field in different geographic locations.

References

Abah, S. O., & Ohimain, E. I. (2011). Healthcare waste management in Nigeria: A case study. *Journal of Public Health and Epidemiology*, *3*(3), 99–110.

Al-Abri, R. (2007). Managing change in healthcare. *Oman Medical Journal*, *22*(3), 9.

Allen, S. (2020). 2020 Global Health Care Outlook: Laying a foundation for the future. *Deloitte.* Available from: https://www2.deloitte.com/global/en/pages/life-sciences-and-healthcare/articles/global-health-care-sector-outlook.html.

American Hospital Association (2015). Sustainability roadmap for hospitals: waste. *Sustainability Roadmap*. Available from: http://www.sustainabilityroadmap.org/topics/waste.shtml#.X0ceg4txXIU.

Azmal, M., Kalhor, R., Dehcheshmeh, N. F., Goharinezhad, S., Heidari, Z. A., & Farzianpour, F. (2014). Going toward green hospital by sustainable healthcare waste management: segregation, treatment and safe disposal. *Health*, *6*(19), 2632.

Bates, D. W. (2000). Using information technology to reduce rates of medication errors in hospitals. *BMJ*, *320*(7237), 788–791.

Bhattacherjee, A., & Hikmet, N. (2007). Physicians' resistance toward healthcare information technology: a theoretical model and empirical test. *European Journal of Information Systems*, *16*(6), 725–737.

Business Wire (2019). The $11.9 trillion global healthcare market: key opportunities & strategies (2014–2022). *ResearchAndMarkets.com*. Available from: https://www.businesswire.com/news/home/20190625005862/en/11.9-Trillion-Global-Healthcare-Market-Key-Opportunities.

Cawich, S. O., Tennant, I. A., McGaw, C. D., Harding, H., Walters, C. A., & Crandon, I. W. (2013). Infection control practice in the operating room: staff adherence to existing policies in a developing country. *The Permanente Journal*, *17*(3), e114.

Chaerul, M., Tanaka, M., & Shekdar, A.V. (2008). Resolving complexities in healthcare waste management: a goal programming approach. *Waste Management & Research*, *26*(3), 217–232.

Chen, H.J., & Chen, W.F. (2003). *U.S. Patent No. 6,568,614.* Washington, DC: U.S. Patent and Trademark Office.

De Regge, M., Gemmel, P., & Meijboom, B. (2019). How operations matters in healthcare standardization. *International Journal of Operations & Production Management*, *39*(9–10), 1144–1165.

deVries, C.G., Hixson, A.N., Liss, B.M., Maddox, W.R., Morin, K., O'Hara, T.E., et al. (2017). *An investigative environmental impact assessment for Kutupalong refugee camp and surroundings, Bangladesh.* University Centre of the Westfjords, Iceland.

Fasola, G., Aprile, G., Marini, L., Follador, A., Mansutti, M., & Miscoria, M. (2014). Drug waste minimization as an effective strategy of cost-containment in oncology. *BMC Health Services Research*, *14*(1), 57.

Gałuszka, A., Migaszewski, Z., & Namieśnik, J. (2013). The 12 principles of green analytical chemistry and the SIGNIFICANCE mnemonic of green analytical practices. *TrAC Trends in Analytical Chemistry*, *50*, 78–84.

Gaynor, M., Li, J., & Vogt, W.B. (2007). Substitution, spending offsets, and prescription drug benefit design. In *Forum for health economics & policy* (Vol. 10, No. 2). De Gruyter.

Gerwig, K. (2005). Waste management & healthcare. Health care without harm. Available from: www.noharm.org/details.cfm. Accessed on 8 November 2020.

Ghasemi, M. K., & Yusuff, R. B. (2016). Advantages and disadvantages of healthcare waste treatment and disposal alternatives: Malaysian scenario. *Polish Journal of Environmental Studies*, *25*(1), 17–25.

Han, X., Man, Z., Xu, S., Cong, L., Wang, Y., Wang, X., et al. (2019). A gold nanocluster chemical tongue sensor array for Alzheimer's disease diagnosis. *Colloids and Surfaces B: Biointerfaces*, *173*, 478–485.

Hasegawa, T., Kondziolka, D., Flickinger, J. C., Germanwala, A., & Lunsford, L. D. (2003). Brain metastases treated with radiosurgery alone: an alternative to whole brain radiotherapy? *Neurosurgery*, *52*(6), 1318–1326.

Healthcare without Harm (2020). Tips of waste reduction. Available from: https://noharm-uscanada.org/issues/us-canada/tips-waste-reduction. Accessed on 7 October 2020.

Helms, M. M., Moore, R., & Ahmadi, M. (2008). Information technology (IT) and the healthcare industry: A SWOT analysis. *International Journal of Healthcare Information Systems and Informatics (IJHISI)*, *3*(1), 75–92.

Hindiyeh, M., Altalafha, T., Al-Naerat, M., Saidan, H., Al-Salaymeh, A., Sbeinati, L. A., et al. (2018). Process modification of pharmaceutical tablet manufacturing operations: an eco-efficiency approach. *Processes*, *6*(2), 15.

Imteaj, A., & Hossain, M.K. (2016). A smartphone based application to improve the health care system of Bangladesh. In *2016 International conference on medical engineering, health informatics and technology (MediTec)* (pp. 1–6). IEEE.

Kalem, G., & Turhan, Ç. (2015). Mobile technology applications in the healthcare industry for disease management and wellness. *Procedia-Social and Behavioral Sciences*, *195*, 2014–2018.

Kjoenniksen, I., Lindbaek, M., & Granas, A. G. (2006). Patients' attitudes towards and experiences of generic drug substitution in Norway. *Pharmacy World and Science*, *28*(5), 284–289.

Lee, C. P., & Shim, J. P. (2007). An exploratory study of radio frequency identification (RFID) adoption in the healthcare industry. *European Journal of Information Systems*, *16*(6), 712–724.

Lluch, M., & Kanavos, P. (2010). Impact of regulation of Community Pharmacies on efficiency, access and equity. Evidence from the UK and Spain. *Health Policy*, *95*(2–3), 245–254.

McHugh, S. M., Corrigan, M. A., Hill, A. D. K., & Humphreys, H. (2014). Surgical attire, practices and their perception in the prevention of surgical site infection. *The Surgeon*, *12*(1), 47–52.

Minoglou, M., Gerassimidou, S., & Komilis, D. (2017). Healthcare waste generation worldwide and its dependence on socio-economic and environmental factors. *Sustainability*, *9*(2), 220.

Moreira, A. M. M., & Günther, W. M. R. (2013). Assessment of medical waste management at a primary health-care center in São Paulo, Brazil. *Waste Management*, *33*(1), 162–167.

Nuralam, H. M., Xiao-lan, Z., Dubey, B. K., & Wen-Chuan, D. (2017). Healthcare waste management practices in Bangladesh: A case study in Dhaka City, Bangladesh. *International Journal of Environmental and Ecological Engineering*, *11*(6), 534–539.

Osmanlioglu, A. E. (2006). Treatment of radioactive liquid waste by sorption on natural zeolite in Turkey. *Journal of Hazardous Materials, 137*(1), 332–335.
Ozder, A., Teker, B., Eker, H. H., Altındis, S., Kocaakman, M., & Karabay, O. (2013). Medical waste management training for healthcare managers-a necessity? *Journal of Environmental Health Science and Engineering, 11*(1), 20.
Padmanabhan, K. K., & Barik, D. (2019). Health hazards of medical waste and its disposal. In *Energy from toxic organic waste for heat and power generation* (pp. 99–118). Woodhead Publishing.
Pinho, S. C., Almeida, M. F., & Nunes, O. C. (2015). Effects of alkaline hydrolysis and autoclaving on inorganic components present in healthcare waste. *International Journal of Environmental Science and Technology, 12*(4), 1191–1200.
Poon, E. G., Jha, A. K., Christino, M., Honour, M. M., Fernandopulle, R., Middleton, B., et al. (2006). Assessing the level of healthcare information technology adoption in the United States: a snapshot. *BMC Medical Informatics and Decision Making, 6*(1), 1.
Rahman, R. (2019). The privatisation of healthcare system in Bangladesh. *International Journal of Health Care Quality Assurance, 32*(1), 97–107.
Reduction in Motion (2020). Sustainability consulting in the healthcare sector: medical waste plan. Reduction in Motion: change the way you think about trash. Available from: https://reductioninmotion.com/sustainability-consulting-healthcare-sector-medical-waste-plan/. Accessed on 9 October 2020.
Rosman, N., Salleh, W. N. W., Mohamed, M. A., Jaafar, J., Ismail, A. F., & Harun, Z. (2018). Hybrid membrane filtration-advanced oxidation processes for removal of pharmaceutical residue. *Journal of Colloid and Interface Science, 532*, 236–260.
Sapkota, B., Gupta, G. K., & Mainali, D. (2014). Impact of intervention on healthcare waste management practices in a tertiary care governmental hospital of Nepal. *BMC Public Health, 14*(1), 1005.
Sharma, P., Kumar, M., Mathur, N., Singh, A., Bhatnagar, P., & Sogani, M. (2013). Health care industries: potential generators of genotoxic waste. *Environmental Science and Pollution Research, 20*(8), 5160–5167.
Singh, R., Kumar, M., Mittal, A., & Mehta, P. K. (2016). Microbial enzymes: industrial progress in 21st century. *3 Biotech, 6*(2), 174.
Subrata, B., Kumar, M. S., Gangopadhyay, A., & Dastidar, S. G. (2013). Materials saving and waste minimization in Indian pharmaceutical industries. *Research Journal of Pharmaceutical Sciences, 2*(2), 20–25.
Thakur, R., Hsu, S. H., & Fontenot, G. (2012). Innovation in healthcare: Issues and future trends. *Journal of Business Research, 65*(4), 562–569.
Thanh, N. P., & Matsui, Y. (2011). Municipal solid waste management in Vietnam: Status and the strategic actions. *International Journal of Environmental Research, 5*(2), 285–296.
Tritter, J., Koivusalo, M., Ollila, E., & Dorfman, P. (2009). *Globalisation, markets and healthcare policy: redrawing the patient as consumer*. Routledge.
Tudor, T. L. (2007). Towards the development of a standardised measurement unit for healthcare waste generation. *Resources, Conservation and Recycling, 50*(3), 319–333.
Tudor, T. L., Noonan, C. L., & Jenkin, L. E. T. (2005). Healthcare waste management: a case study from the National Health Service in Cornwall, United Kingdom. *Waste Management, 25*(6), 606–615.
Tudor, T. L., Woolridge, A. C., Bates, M. P., Phillips, P. S., Butler, S., & Jones, K. (2008). Utilizing a systems' approach to improve the management of waste from healthcare facilities: best practice case studies from England and Wales. *Waste Management & Research, 26*(3), 233–240.
UNEP & WHO (2002). Fundamentals of healthcare waste management. *Safe management of health-care waste*. World Health Organization. Available from: https://www.who.int/water_sanitation_health/medicalwaste/en/guidancemanual1.pdf.

UNEP (2020). Healthcare waste: what to do with it? *unenvironment.org*. Available from: https://www.unenvironment.org/news-and-stories/story/healthcare-waste-what-do-it.

Velagaleti, R., & Burns, P. (2007). A review of the industrial ecology of particulate pharmaceuticals and waste minimization approaches. *Particulate Science and Technology*, *25*(2), 117–127.

WHO (2005). *Management of solid health-care waste at primary health-care centres: A decision-making guide*. Available from: https://www.who.int/water_sanitation_health/medicalwaste/decisionmguide_rev_oct06.pdf?ua=1.

WHO (2018). *Health-care waste*. World Health Organization, Geneva. Available from: https://www.who.int/news-room/fact-sheets/detail/health-care-waste.

Yang, C., Peijun, L., Lupi, C., Yangzhao, S., Diandou, X., Qian, F., et al. (2009). Sustainable management measures for healthcare waste in China. *Waste Management*, *29*(6), 1996–2004.

CHAPTER 5

Source reduction and waste minimization in construction industry

5.1 Introduction

The state of the construction industry is one of the most important components of any country's economy. It is because, construction industry is a determinant of the country's technological and technical advancement, often regulating the growth of the country's infrastructural development that often directs to the country's advancement in terms of sustainability assurance. Construction industry refers to the industrial branch of manufacturing and trade related to building, repairing, renovating, and maintaining infrastructures (Sustainable Development Indicator Group, 1996). Unlike the recent progress of other industries, the concept of construction industry emerged long time back, nearly before 3000 years ago among Egyptian civilization, since they had been one of the very first civilizations who organized and built large scale infrastructures using bricks and stones (Vault, 2020). Being a very old industry, construction industry has undergone a lot more evolutions compared with many other industries. The history of construction industry experienced a sudden shift since the industrial revolution started, as manufacturing and trading became easier since then. The growth of construction industry flourished even more during the colonialism period. There are evidences of constructing 14–15 storied buildings before last century (Vault, 2020). Co-current advancements related with construction industry involve incorporation of new materials, methods, and technologies. Along with these, sustainability concerns are also involved with construction industry nowadays. All these refer to the huge scope of improvements in this sector.

However, construction industry doesn't merely refer to only the construction process of infrastructures. Instead, it incorporates all the corresponding manufacturing, renovating, and trading opportunities associated with infrastructure construction. The extents of this industry can be classified under several classes based on several criteria. For example, in terms of

Source Reduction and Waste Minimization
http://dx.doi.org/10.1016/B978-0-12-824320-6.00005-8

the ownership of the constructed infrastructures, construction can be classified as, public or private. Private construction projects are those that are constructed for private usage of the owners and the construction work is also funded from private funds. On the contrary, public constructions are mostly constructed for public welfare that are owned and funded by public funds, usually collected through tax, revenue, and others. Besides, based on the purpose of the constructed infrastructures, construction can be classified as residential and non-residential construction. Residential construction incorporates manufacturing, constructing, and trading involved with residential infrastructure construction such as housings, garages, etc. On the other hand, non-residential construction involves a variety of non-residential infrastructure construction such as bridges, commercial buildings, hospitals, schools, roads, and so on. Understanding the difference in between residential and non-residential construction is important, since they have several differences in terms of the corresponding construction mechanisms adopted.

In addition to the criteria set as per the ownership and purpose of construction, Census.gov (2020) categorizes construction industry under classes such as heavy construction, construction by special trade contractors, and land development-buildings construction. This classification system classifies different commonly adopted construction mechanisms that often determine the quality, durability, and sustainability of the construction project. Here, heavy construction denotes construction mechanism where all the responsibilities are assigned upon one single entity including sales, marketing, constructing, and others. Then comes special trade contractor initiated construction where specialized construction related activities are assigned to certain dealers or contractors under subcontract mechanism or under the direct supervision of the primary contractors. And land development-buildings construction is the most common and general construction mechanism where the owner hires commercial contractors and assign all the responsibilities associated with infrastructure construction, from land subdivision and land development to building infrastructures. Thus, the construction industries are regulated under different mechanisms (Census.gov, 2020). Though the suitability of different construction mechanisms vary, altogether such industrial initiatives contribute to the country's growth.

Construction industry plays an important role in the country's economic growth. Even there are a lot of scopes and opportunities to improve the status of the construction industry; thus, contributing to the national growth. It is because, day by day the urge to establish more improved, durable, and sustainable construction procedures are being encouraged. Therefore, it is yet growing industry since newer demands and expectations are being incorporated for infrastructure construction. Realizing the future potentials of this industry, the global construction industry is estimated to reach 10.5 trillion USD worth industry by 2023. And it is expected that the construction industry will experience a compound annual growth rate (CAGR) of 4.2% in between 2018 and 2023 (Research and Markets, 2020). This depicts the significant role that the construction industry may play toward the economic growth of the country.

However, construction industry is also found to be responsible for good share of waste generation. Since most of the construction initiatives are held in the areas of huge public crowd, the inability to ensure suitable source-level waste minimization initiatives of construction activities will lead to large scale exposure and significantly negative environmental impacts. Besides, numerous number of role players are acting in this industrial sector which makes it even more difficult to co-ordinate and regulate among these stakeholders. Along with these, since most of the construction works are held within a short duration, there are also very limited scopes to design and implement long-term source-level waste minimization initiatives. Thus, all these different challenges make incorporation of source-level waste minimization approaches even more difficult.

This chapter explores some of the commonly produced waste in the construction industries along with their corresponding construction sources and hazards. And it also suggests approaches to ensure source-level minimization of such waste produced by construction industries.

5.2 Commonly produced waste in the construction industries: source and biogeochemical hazards

The types of generated waste in the construction industries are diverse in nature. As per 2019, only Australia produce 20 Mega tons of construction waste is generated annually, most of which anyhow reach to landfill deposition (Ketchell, 2019). Being the biggest natural resource consuming industry of the United Kingdom, the construction industry produces 100 million tons of construction waste annually which is one-fourth of the total material consumption by the construction industry and over 50% of the total generated waste in the United Kingdom (PBC News, 2019). This shows how inefficient the construction industry is in terms of resource consumption and waste generation; even in a developed country where there are adequate scopes of source-level waste minimization. The situation will get even worse in near future as per the prediction of Transparency Market

Research, since they estimate that the amount of such construction industry produced waste will be doubled by 2025 (Slowey, 2018). Unfortunately, this vast share of construction waste is so diverse in nature that categorizing these wastes into suitable classes is a quite difficult task. Most of such wastes are generated from demolition, roadwork, and excavation causing production of metallic, non-metallic, liquid, gaseous, and many more types of pollutants (Slowey, 2018). Therefore, instead of categorizing the construction industry produced waste into broader waste classes, some of the commonly produced wastes in the construction industries have been introduced in Table 5.1 along with their corresponding construction sources and hazards.

5.3 Source-level waste minimization initiatives by the construction countries

As mentioned in Table 5.1, construction industry may produce a variety of waste throughout different construction activities. Since most of the construction activities occur nearby human settlement, there are chances of easy contact of these wastes with human when such wastes are exposed to the environment. Therefore, it is urgent to prevent production of this large amount of construction waste to put an end to its improper exposure. In this regard, source-level minimization of these construction wastes can be preferred over the treatment of construction waste, since such source-level waste minimization will prevent the waste to be exposed to the environment. In the process of waste source reduction, the waste is being prevented to be generated on-site; thus, it stops its way toward open disposal. Such source level waste minimization can be preferred over other conventional construction waste minimization initiatives such as landfilling.

Landfilling can be regarded as the most common traditional construction waste minimization strategies. The total landfill waste of 32% comes from construction waste and 13% of the purchased products for construction activities are directly sent to landfills without being used at all (Designing Buildings Wiki, 2020). Despite being the most popular construction waste minimization system, landfilling is a significantly unsustainable process, as it puts a threat toward the environmental sustainability. Landfill cause carbon containing GHG emission in long-term; thus, contributes to global warming and climate change (Scharff, 2014). Landfilling practice is also responsible for leaching of harmful chemicals causing soil and groundwater pollution. Especially when construction wastes are sent to the landfill sites, the exposure pathways increases throughout the transportation and

Table 5.1 Commonly produced waste in the construction industries and corresponding hazards (Slowey, 2018; Designing Buildings Wiki, 2020; PBC News, 2019; Muleski et al., 2005; Araújo et al., 2014).

Waste	Construction sources and description of the waste	Hazards
Metallic waste	Metallic waste denotes waste that contains different metallic components. Though metallic waste generated in the construction industries are not that diverse in nature comparing with the electrical and electronics industries, there are a good amount of metallic waste generated from construction activities. Cables, pipes, wires, etc. are the common metallic waste generation source for construction industries. Besides, during the process of infrastructure construction several metallic instruments, machineries, and technologies are used that also produce good amount of metallic waste. Metallic waste produced by the construction industries mostly contains iron, steel, alloys, and some other commonly found metals.	• Long-term exposure to metallic waste leads to different health consequences which are metal specific; for example, exposure to iron may lead to anemia, exposure to nickel leads to skin irritation. • Metallic waste leaches through soil or can easily get mixed in the aquatic environment causing death of useful microbes, degraded environmental conditions, and other consequences.
Plastics	Plastics are synthetic organic polymers produced from a variety of polymeric substances such as polyvinylchloride, polyethylene, nylons, and others. Plastics wastes are a global concern nowadays due to their excessively high degradation time. Like other industries, construction industries are also responsible for good amount of plastic waste generation that may sustain in the environment for a long period of time. Construction industry produced plastic wastes include garbage bags, carry bags, disposable/one-time materials, packaging of construction materials, and others. Additionally, machineries used for construction works also cause generation of good share of plastic waste.	• Plastic pollution significantly hampers aquatic ecosystem putting the life of the aquatic species in danger. • Improper management of plastic waste (i.e., incineration, landfilling, etc.) emit huge amount of carbon dioxide, methane, carbon monoxides, and other GHGs; thus contribute to global warming. • Exposure to micro-plastic significantly degrades human health conditions.

(Continued)

Table 5.1 Commonly produced waste in the construction industries and corresponding hazards (Slowey, 2018; Designing Buildings Wiki, 2020; PBC News, 2019; Muleski et al., 2005; Araújo et al., 2014). (*Cont.*)

Waste	Construction sources and description of the waste	Hazards
Bricks	Bricks are one of the most common entities used during construction work. After the destruction of infrastructures or during the construction of infrastructures, a good amount of bricks are left as waste. Therefore, in many cases, bricks are regarded as the most common waste produced by construction industries. Bricks are mostly used to provide the structural appearance of the infrastructures. Bricks often contain a good share of metallic waste such as calcium, magnesium, aluminum, and others.	• Improper disposal of bricks leads to emission of good amount of metal waste and also cause carbon emission in the environment. • Bricks can be easily recycled and reused for which disposing bricks is a very inefficient resource utilization mechanism.
Ceramic	Ceramics are chemical components that contain good amount of Silicon dioxide along with other metallic oxides. Construction industries produce huge amount of ceramic waste annually, since ceramics are commonly used as a very durable and strong element. Especially for electricity supply system installation, basins, commodes, and other utilities, ceramics are widely used in the construction industry.	• Metallic pollutant exposure from ceramics leads to metal-borne biogeochemical hazards on the environment and on human health. • Improper management of ceramics leads to physical hazards causing bleeding, scratching, and others.
Cardboards/ woods	Both conventional and modern construction require to use a lot of woods and cardboards, since these are highly sustainable and easily replaceable options; especially for interior designing, artistic beauty enhancement, and for other purposes cardboards and woods are often used. However, since these are easily perishable, frequent changing of cardboards and woods are often needed in the construction industries. It causes generation of a good amount of cardboard/wooden waste.	• There is not many hazards that cardboards or wooden waste may cause. However, improper management of these wastes (i.e., landfills, incineration, etc.) may lead to carbon emission contributing to global warming and climate change indirectly.

Shingles/ stones	Shingles and stones are often used during construction activities which might be released as construction waste in the form of large pieces or as crushed particulate form. Whatever the form is for these wastes, these can cause different sort of pollution being released in the environment.	• Improper management of shingles or stones may cause physical hazards or may lead to soil/ water pollution.
Concrete/ cement/ glass/ gypsum	Concrete, cement, glass, and gypsum—all these are different chemical waste generated commonly by the construction industry. These are needed for construction activities and usually the extra amount of these chemicals are dumped in open space after the construction activities. Even though these chemicals do not contain much of the harmful ingredients, a large amount of such chemical waste are misused and released almost after every construction activities. Therefore, proper management of these wastes are needed.	• Exposure of these chemical wastes into the environment degrades soil and water parameters and kills useful microbes and aquatic species. • Though direct exposure to these, do not cause much health concern in human, however, if it enters into the food chain, there will be extremely negative impacts on human health.
Paints, varnishes	Paints and varnishes are liquid form of chemical waste commonly produced during construction activities. Even after the construction, these types of liquid waste can be generated constantly from the infrastructures during renovation and repairing of the infrastructures.	• Being liquid waste materials, paints and varnishes can easily get mixed with the aquatic or soil ecosystem causing negative impacts on the useful microbes and aquatic species.
Air pollutants	Construction industry produces a good amount of air pollutants in the form of dusts and particulates. Especially the particulate matters emitted during construction activities can be considered as truly harmful air pollutants. During construction of infrastructures, large pieces of components are crushed and converted into smaller component of dirt, dust, and other harmful chemicals. Because of being components of very small size, these particles travel a very long distance and the smaller weight of the particles does not let them to be destroyed so easily. Thus, these TSPs (total suspended particulates) and PMs (particulate matters) sustain in the environment for a long time putting a threat on the environment causing air pollution.	• These air pollutants increase the temperature of our environment and contribute in global warming. • These matters enter in our respiratory system very easily and may initiative respiratory diseases, heart/lungs dysfunction, breathing problem, and many other health concerns. • PMs of harmful chemicals can also irritate sensitive organs (i.e., eyes, skin, etc.) • Hampers vegetation growth reducing light absorption for photosynthesis.

dumping process. Besides, considering all the impacts and consequences of landfill sites, it is a more expensive waste management initiative compared to many other advanced waste treatment technologies (Designing Buildings Wiki, 2020). Especially in the developed countries, frequent landfilling cause to spend a good amount of money in the form of landfill taxation and other regulations; at the same time it also reduces the property prices sometimes (Scharff, 2014; Seok Lim and Missios, 2007). Considering all these different negative impacts that landfilling of construction waste may bring about, source-level waste minimization can be regarded as the most suitable alternative in terms of the safety, sustainability, and long-term cost effectiveness in the construction industries.

However, construction industry doesn't have many different waste source reduction alternatives to be implemented, since implementing on-site waste minimization initiatives for the short duration of construction initiatives is quite challenging. Instead, a collaborative planned approach is needed to assure minimization of construction waste generation. Adopting immediate post generation sustainable biochemical and thermal treatment strategies is tough in this particular industrial context, since on-site installation of the facilities is a tough, time-consuming, and inefficient task. It is because the treatment technology will be of no use when the construction work will come to an end and at the same time, transferring the installed treatment facilities is also an expensive task. This is why source-level minimization of construction waste should focus more on process and system modifications bringing changes in the administrative and process mechanisms. This is why in this chapter strategies discussed for source-level minimization of construction waste have mostly covered process modifications, 3R approach implementation, and policies interventions instead of prioritizing immediate postgeneration treatment (Fig. 5.1).

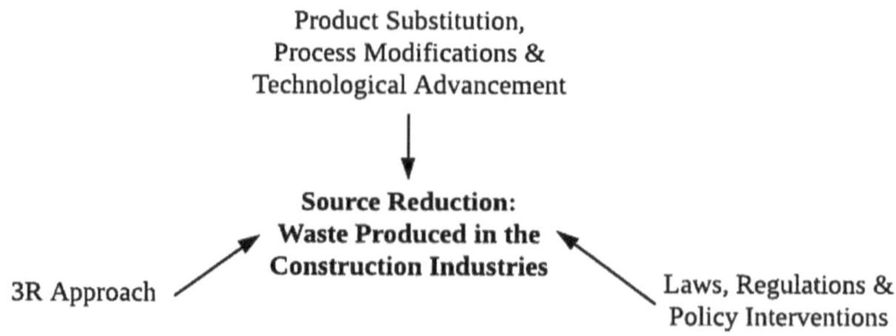

Figure 5.1 ***Overview on the waste source reduction mechanism for the construction industries.***

5.3.1 Product substitution, process modifications, and technological advancement

Process modification can be considered as one of the most suitable source-level waste minimization initiatives, since it can prevent pregeneration level waste minimization leading to least amount of hazards. In case of construction industry, process modification mechanism with an aim to source-level waste minimization can be considered fully as an industrial responsibility, since imposing process modification initiatives through regulation is tougher where numerous numbers of smaller players are involved in a small industry like construction industry. Besides, increase in the amount of waste generation also increases the corresponding waste management cost for the industrial role players of the construction industries. Along with these, since Table 5.1 illustrates some of the major corresponding hazards that construction waste might bring about, for the safety of human health and protection of our environment, source-level waste minimization initiatives through process modifications may play important roles minimizing the pathways of construction waste generation (Ketchell, 2019). Process modification strategy can be as simple as efficiency enhancement to as complex as mobile waste treatment plant set-up in the construction site. Whatever complexity or simplicity the process modification initiative incorporates, understanding the dreadful consequences construction waste may bring about and also to ensure sustainability of the construction industry, it is necessary to adopt less waste producing processes by the construction industries.

Among many of the process modification initiatives, product substitution can be regarded as the primary approach to ensure either less amount of waste producing or less toxic waste producing processes. For example, a lot of innovations are going on currently to determine more sustainable infrastructure construction materials considering the long-term sustainability potentials of the materials and the infrastructures. Even a good number of initiatives are going on currently that aims infrastructure construction from food waste such as corn cobs, wheat, banana plants, and others. Along with these, initiatives are going on to substitute leather and fabrics usage by more sustainable products (PBC News, 2019). If the conventionally used construction materials are substituted by such comparatively more sustainable and biodegradable products, this will help to substitute metallic or chemical waste produced during construction activities. Product substitution with an aim to source-level waste minimization should also incorporate proper standardization as per the geological, environmental, and socio-economic context of the corresponding construction industry (Hills, 2017). Only focusing on the waste minimization potentials through product substitution initiatives instead of checking its suitability, durability, and effectivity properly may cause unsustainability of the constructed infrastructures.

Product substitution initiatives can also be expanded through careful planning by the construction industrialists. New innovations and newly discovered materials should be incorporated in the construction initiatives. For instance, steel formwork system can be incorporated to replace concrete works. Steel formwork can be a more sustainable system compared with concrete works, since steel formwork can be reused later for other construction activities thus ensures minimization of waste generation even after end-use (Designing Buildings Wiki, 2020). Besides, since wooden frameworks are comparatively a safer option in terms of corresponding waste generation potentials, wooden components can be introduced for substituting metallic frameworks. This can be more appropriate for interior or exterior designing of constructed infrastructure. Such initiatives for product substitution initiatives should carefully monitor and assess material efficiency comparison among all the suitable alternatives (Sharman, 2018). Material efficiency assessment should explore sustainability, cost-efficiency, recycling-reuse potentials, waste generation potentials, and many other features so that considering all these different criteria, the most suitable alternative can be chosen. Unless enough monitoring, evaluation and impact assessment is conducted before introducing product substitution, there will be several negative outcomes instead of getting the expected sustainable outcomes.

Along with suitable product substitution, dependency on modern technologies can be used as a process modification tool to ensure source-level waste minimization. Since the ever growing co-current trend of urbanization requires frequent construction of infrastructure incorporating modern technologies and newer innovations, technologies are constantly being upgraded incorporating newer tools and innovative features. Even in the construction sector, modern technologies are being incorporated at a daily basis to assure better facilities. In all different industries, technological advancement is often regarded as a process modification initiative to assure better resource efficiency (Dong et al., 2017). Especially in case of construction industry, technological advancement might be considered as a tool of improving process efficiency as well as resource efficiency. However, for sustainable source-level waste minimization, it is needed to properly assess the impacts that such technologies may bring about. The construction industrial authorities should play an important role in adopting such technologies, since it will minimize their burden of construction waste generation and at the same time, it will being about many other co-benefits at the same time such as cost efficiency, increases durability/sustainability and others (Machado and Rodrigues, 2018). Thus, even with the slightest willingness of the industrial authorities to assure suitable technology adoption, it will be possible to bring about significant positive outcomes in source-level waste minimization in the construction industries.

Technological interventions as a process modification initiative for construction industries can be applicable in different scenarios. For example, construction waste minimization initiatives in the United Kingdom is significantly dependent on the emergence of newer technologies adopted by all the stakeholders related with the construction industry (Osmani, 2012). A survey-based research conducted in multiple construction sites of the UAE also found the willingness and urgency to expand technological interventions in managing the construction materials and processes more efficiently (Al-Hajj and Hamani, 2011). Also, another study conducted in several Sri Lankan construction sites found association in between different technological initiatives and material wastage (De Silva and Vithana, 2008). Thus, choosing the most suitable processes or technologies from different options of materials and technologies may contribute in reducing the total generated waste from construction industry. Though many of the industrialists deny such systematic investigation based technological alteration, the long-term benefits that it may bring about will compensate the capital cost spent for such systematic technological innovations used in the construction industries (Gulghane and Khandve, 2015). Thus, process modification initiatives can be introduced in different sectors of the construction industries through product substitution and technological alteration.

5.3.2 3R approach

3R approach for source-level construction waste minimization incorporates systematic reduction of waste through reduction, reuse, and recycling. Unlike some other industries, there are tremendous opportunities of 3R approach implementation in case of the construction industries, since most of the waste generated in this industry do not degrade or decay easily for which they can used multiple times after slight modifications. Therefore, reduction, reuse, and recycling can be considered as an easier source-level waste minimization initiative in this regard. Implementing 3R approach is a very sustainable waste minimization approach, as it doesn't incorporate much complexity and ensure resource efficiency, cost savings, and many more co-benefits at the same time (Aadal et al., 2013). Considering all these different benefits of 3R approach implementation in construction waste minimization, many countries are adopting large-scale initiatives to reduce, reuse, and recycle construction waste. In this regard, China can be a good example. The total generated waste of 30%–40% in China comes from construction industry. Therefore, they are trying to adopt intense 3R approach implementation through innovative technology adoption, market model designing, and through economic incentives offered for 3R approach implementation in the construction industries (Huang et al., 2018). These initiatives can also be applicable for 3R approach mediated construction waste minimization in other geographic context as well.

Reduction being the primary step in this 3R approach can be regarded as initiatives that will reduce consumption and misuse of resources. A large amount of total generate construction waste comes from carelessly caused damaged, loss and over-purchasing; for example, at least 10% of construction materials in the United Kingdom are wasted in these ways (Osmani, 2011). This percentage is even higher for the developing countries due to their limitations of investment in proper planning. This vast amount of resource misuse is not only regarded as an example of inefficient resource consumption but also a major cause of monetary loss and environmental degradation. Only some planned initiatives as per the reduction strategies of 3R approach can end such huge loss of resources. This shows the urgency to reduce material purchase and wastage. In case of the construction industry, the materials purchased should be reduced and misuse of materials should also be strictly prohibited thorough regulation. It not only causes a huge monetary loss but also this misuse is one of the main contributors to such large amount of unexpected but avoidable construction waste generation (PBC News, 2019).

In addition to reduction, suitable strategies for reusing construction materials should also be introduced in the construction industries. Vast amount of reusable construction waste is produced every year; most of which are improperly managed or dumped into the landfill sites. This includes all the metallic waste, bricks, woods, and other waste materials. These wastes can be easily used for other construction works without any complex recycling initiatives. Though it is true that the performance of the materials may degrade due to repeated use, if proper management strategies are introduced under a centralized or semi-centralized system, this performance can be improved tremendously (Sivakumar et al., 2004). For example, when any infrastructure is destroyed the potential waste can be sold to others under a circular economy system where resources flow in a society efficiently. In doing so, on-site waste segregation or waste classes sorting need to be assured; otherwise there are chances of more misuse with less reuse potentials of resources (Wang et al., 2010). Reuse can thus be regarded as a quite successful source-level waste minimization initiative.

Along with reduction and reuse, recycling opportunities in the construction industry is also quite vast. It is because most of the generated construction wastes are usually of quite good condition due to their extremely high durability, it is possible to slightly recycle and reuse many of such waste. Since there are good opportunities of recycling of construction waste, material recycling should be encouraged. It will not only protect our environment from waste accumulation but also save up money to purchase newer materials and reduce landfill taxation (PBC News, 2019). In addition to it, construction waste recycling-based start-ups will also rise, if the industrialists are not willing to take direct recycling initiatives. Thus, a new sustainable business framework will also be formed. For example, since good amount of bricks are generated as construction waste often, they can be crushed and reused for some other construction activities. Aside from these, as a significant share of chemical waste such as metals, glass, cement, alloys, etc. are produced from construction activities, these can also be recycled to be utilized in construction industry or in some other industries. The contractors should also be trained and encouraged to undertake recycling initiatives for a more successful 3R approach implementation in the construction industries (Ng et al., 2015). For instance, in Malaysia, the contractors who have received proper training and understood the urgency and their own roles and benefits out of 3R approach implementation in corresponding construction projects (Ng et al., 2017). This is how the role of the stakeholders in a bottom-up approach can promote recycling initiatives in the construction industries.

5.3.3 Laws, regulations, and policies interventions

Like all other industries, successful source-level waste minimization of construction industry also requires interventions of the government and other regulating agencies. The government and regulating agencies should take the initiatives to increase the practices and knowledge of the role players from the construction industries. Introducing good practices of rundown approach can be a potential solution in this regard, since this will make sure at least 15% more on-site waste reduction and up to 43% less landfill disposal of construction waste (PBC News, 2019). In this process of run-down approach, a planning and need-based approach is adopted that includes several steps such as predesigning, designing and procurement, preconstruction initiatives, during construction initiatives, and finally postconstruction initiatives. All these different steps makes it easier to reduce significant share of construction waste. This is part of site-waste management plan which should be legally made compulsory for all the construction activities (Designing Buildings Wiki, 2020). This is why and how, the government and regulating agencies should promote planned initiatives introducing new laws, policies, and other forms of interventions.

Along with planning, there are other initiatives such as standardization, gap minimization, enhanced preparedness, and others which should be introduced in this regard. All different industrial setting often require standardization of chemical and physical standards. Especially in case of a vast industry like construction industry, there are more associated complexities due to the large amount of produced waste and contribution of a lot of smaller role players in the industry. Standardization often plays an important role to encourage resource recovery and waste minimization initiatives. For example, The Minnesota Pollution Control Agency (MPCA) imposed tougher standards for demolition landfills to encourage resource recovery technologies and to minimize groundwater and soil contamination from landfill disposal of construction waste (Slowey, 2018). Another challenge that the regulating agencies should tackle is the gaps among different stakeholders. At the first place, the gaps in between the policymakers and the contractors should be minimized in this regard (Ng et al., 2015). Even the media coverage should be increased so that stakeholders of all different classes are acknowledged about their corresponding roles and responsibilities to assure a more sophisticated source-level waste minimization initiative. Preparedness of the government should be enhanced to tackle natural hazards or any other anthropogenic hazards that might be initiated due to the improper management of construction waste. Besides, educational and training initiatives should also be introduced for the construction industrial role players. Such training opportunities should be expanded and made compulsory through the government regulations.

Case Study 6: Reconomy's Effort to Incorporate Waste Management Initiatives in Business Sectors

Reconomy is a business initiative that aims to inspire sustainable, smarter, and greener business practices. They assist different industries to design and contextualize waste minimization initiatives. They envisions to create sustainable environment through smarter and cleaner practices and in doing so, they provide consultancy, technical, and technological support. They have also provided waste minimization support to different construction activities with an aim to promote sustainable, durable, and environment-friendly construction activities. Reconomy provides construction supports under different steps of mechanisms that include minimizing landfill costs, segregating waste efficiently, maximizing waste's potentials, reusing existing materials, and finally allying with the right suppliers (Reconomy, 2017). Reconomy has been providing sustainable construction support for over 25 years incorporating up to date cleaner and newer technologies. However, the example of Reconomy can be inspiring for entrepreneurs with relevant expertise who can launch similar type of initiatives under sustainable business framework. They should also be promoted by the industrial regulating agencies so that the contractors get motivated to adopt services of such firms that offer suitable contextualize construction waste minimization initiatives.

References

Aadal, H., Rad, K. G., Fard, A. B., Sabet, P. G. P., & Harirchian, E. (2013). Implementing 3R concept in construction waste management at construction site. *Journal of Applied Environmental and Biological Sciences*, *3*(10), 160–166.

Al-Hajj, A., & Hamani, K. (2011). Material waste in the UAE construction industry: Main causes and minimization practices. *Architectural Engineering and Design Management*, 7(4), 221–235.

Araújo, I. P., Costa, D. B., & De Moraes, R. J. (2014). Identification and characterization of particulate matter concentrations at construction jobsites. *Sustainability*, *6*(11), 7666–7688.

Census.gov (2020). North American Industry Classification System. *Census.gov.* Available from: https://www.census.gov/eos/www/naics/reference_files_tools/1997/sec23.htm.

De Silva, N., & Vithana, S.B.K.H. (2008). Use of PC elements for waste minimization in the Sri Lankan construction industry. *Structural Survey*, *26*(3), 188–198.

Designing Buildings Wiki (2020). *Construction waste.* Available from: https://www.designingbuildings.co.uk/wiki/Construction_waste.

Dong, L., Dai, M., Liang, H., Zhang, N., Mancheri, N., Ren, J., et al. (2017). Material flows and resource productivity in China, South Korea and Japan from 1970 to 2008: A transitional perspective. *Journal of Cleaner Production*, *141*, 1164–1177.

Gulghane, A. A., & Khandve, P.V. (2015). Management for construction materials and control of construction waste in construction industry: a review. *International Journal of Engineering Research and Applications*, *5*(4), 59–64.

Hills, R. (2017). Substitution of Construction Products: A Guide to Managing Product Substitution. *Building Products Innovation Council (BPIC).*

Huang, B., Wang, X., Kua, H., Geng, Y., Bleischwitz, R., & Ren, J. (2018). Construction and demolition waste management in China through the 3R principle. *Resources, Conservation and Recycling*, *129*, 36–44.

Ketchell, M. (2019). We create 20m tons of construction industry waste each year. Here's how to stop it going to landfill. *The Conversation*. Available from: https://theconversation.com/we-create-20m-tons-of-construction-industry-waste-each-year-heres-how-to-stop-it-going-to-landfill-114602.

Machado, C. A., & Rodrigues, S. C. (2018). Environmental contamination by technogenic deposits in the urban area of Araguaína, Brazil. In *Urban geomorphology* (pp. 114–125). Elsevier.

Muleski, G. E., Cowherd, C., Jr., & Kinsey, J. S. (2005). Particulate emissions from construction activities. *Journal of the Air & Waste Management Association*, *55*(6), 772–783.

Ng, L. S., Seow, T. W., & Goh, K. C. (2015). Implementation on solid waste reduction through 3R (NSWM Policy) and elements to close gap between policy and contractors in construction industry in Penang. *International Journal of Environmental Science and Development*, *6*(9), 668–675.

Ng, L.S., Tan, L.W., & Seow, T.W. (2017). Current practices of construction waste reduction through 3R practice among contractors in Malaysia: Case study in Penang. In *IOP Conference Series: Materials Science and Engineering* (Vol. 271, No. 1, p. 012039). IOP Publishing.

Osmani, M. (2011). Construction waste. In *Waste* (pp. 207–218). Academic Press.

Osmani, M. (2012). Construction waste minimization in the UK: current pressures for change and approaches. *Procedia-Social and Behavioral Sciences*, *40*, 37–40.

PBC News (2019). How is the construction sector combatting its waste. *The PBCToday*. Available from: https://www.pbctoday.co.uk/news/planning-construction-news/waste-in-construction/65702/.

Reconomy (2017). Top 5 practical waste-saving tips for your construction site. *Reconomy*. Available from: https://www.reconomy.com/2017/04/04/top-5-practical-waste-saving-tips-for-your-construction-site/.

Research and Markets (2020). Growth opportunities in the global construction industry. *Research and Markets*. Available from: https://www.researchandmarkets.com/reports/4439921/growth-opportunities-in-the-global-construction.

Scharff, H. (2014). Landfill reduction experience in The Netherlands. *Waste management*, *34*(11), 2218–2224.

Seok Lim, J., & Missios, P. (2007). Does size really matter? Landfill scale impacts on property values. *Applied Economics Letters*, *14*(10), 719–723.

Sharman, J. (2018). Construction waste and materials efficiency. *The NBS*. Available from: https://www.thenbs.com/knowledge/construction-waste-and-materials-efficiency.

Sivakumar, V., McKinley, J. D., & Ferguson, D. (2004). Reuse of construction waste: performance under repeated loading. *Proceedings of the Institution of Civil Engineers-Geotechnical Engineering*, *157*(2), 91–96.

Slowey, K. (2018). Report: Global construction waste will almost double by 2025. *Construction Dive*. Available from: https://www.constructiondive.com/news/report-global-construction-waste-will-almost-double-by-2025/518874/.

Sustainable Development Indicator Group (1996). Construction industry. *Working Draft Framework, 2*. Available from: https://www.hq.nasa.gov/iwgsdi/Construction.html. This includes drilling and solid mineral exploration.

Vault (2020). *Research industries: construction*. Available from: https://www.vault.com/industries-professions/industries/construction.

Wang, J., Yuan, H., Kang, X., & Lu, W. (2010). Critical success factors for on-site sorting of construction waste: a China study. *Resources, Conservation and Recycling*, *54*(11), 931–936.

CHAPTER 6

Source reduction and waste minimization in chemical industry

6.1 Introduction

Chemical industry is a vast industry that incorporates all different types of product producing industries whose generation is based on heavy use of chemicals. Usually, industries that are involved with industrial chemical generation are broadly known as chemical industry. As per the end-products, processes or raw materials associated with chemical industries, such industry can be classified into several classes. Understanding the huge diversity of chemical producing industries, such chemical industries are usually categorized under broad categories such as inorganic and organic chemical producing industries, fertilizer industries, refining and petroleum industries, electroplating and heat treatment industries, pesticide industries, hydro generated oil/soap industries, and so on (Soni, 2020). Along with these, sometimes pharmaceutical industries, agricultural product processing industries, chemical processing/supplying industries, etc. are also regarded as specialized chemical industries. Thus, chemical industry covers a variety of industries which are focused on producing chemicals of different purposes. Understanding these varieties of classes under chemical industry is important, since such clustering helps to design suitable policies, initiatives, and interventions more readily (Banerjee et al., 2003). The variety and extent of such chemical industries are more diverse especially in the developed countries where there are adequate opportunities for industrial growth (Arora et al., 2001). The developing countries are not much focused on the growth of their chemical industries since they are already struggle to meet the basic demands of their citizen and to maintain a sustainable economic condition.

All these different types of industrial inclusions make chemical industry a vast industry incorporating investment and supply chain management of multiple different types of industries. This is why investment and compound annual growth rate (CAGR) of the chemical industries are often higher than many other industries. The commonly considered subcategories of chemical industry is expected to reach to a worth of

Source Reduction and Waste Minimization
http://dx.doi.org/10.1016/B978-0-12-824320-6.00006-X

953.9 billion of USD by 2027 at a 5% CAGR from the 873.38 billion of USD worth market of 2019 (Allied Market Research, 2020; BCC Research, 2020). Though, the growth percentage might not seem that high, its corresponding monetary value of growth is extremely high and can regulate the growth of the country's economy tremendously. Even in 2016 and 2017, the CAGR of the chemical industry's total performance shareholder returns (TSR) experienced over 24% CAGR (McKinsey). This percentage of CAGR is significantly high considering the wide extent of multiple sectors of chemical industries. Unfortunately, many of such chemical industries are not much focused on their environmental sustainability; instead they are more focused on profit making. In this regard, Domazlicky & Weber (2004) argue that economic growth without enough initiatives of environmental protection won't be sustainable and such unsustainability will lead to comparatively higher expenditure in the long run.

Being one of the largest cluster of different types of product producing industries, chemical industries cause generation of different types of waste, most of which are highly hazardous chemical wastes and such wastes are responsible for causing environmental unsustainability. It is because most of such chemical wastes are not biodegradable and contain highly toxic chemicals. Besides, since such chemical industry produced wastes are generated under certain extreme physical or chemical conditions (i.e., high pressure, high temperature, concentrated acidic, or alkaline conditions, etc.), handling such waste sustainably is a huge concern. Improper handling and maintenance of these highly toxic wastes may bring about degradation of health or environmental conditions. Additionally, proper management or treatment of chemical wastes is also quite expensive, as such management initiatives often require sophisticated technological and technical set-up. Considering corresponding economic, technical, and hazard-related concerns of chemical industry produced waste, pregeneration waste minimization, known as source-level waste minimization, can be preferred over postgeneration treatment or management initiatives for chemical industry produced waste (Chaaban, 2001). Thus, source reduction and waste minimization approaches can be considered as source level chemical waste minimization initiatives.

This chapter explores and categorizes some of the major classes of waste produced in different chemical industries. At the same time, it discusses on some of the major source-level waste minimization initiatives that can be adopted by the chemical industries.

6.2 Commonly produced waste in the chemical industries: source and biogeochemical hazards

The extent and variety of waste produced in different chemical industries are so diverse that categorizing these wastes into suitable classes in a major challenge. However, since all the different industries produce different chemicals as their major waste, this chapter will discuss some of the major chemical industry produced waste under three classes: solid, liquid, and gaseous waste. Though classifying waste as per their physical properties is not much relatable to understand the toxicity and corresponding hazards of the waste, however, physical property based classification of chemical waste helps to understand the variety and extent of the chemical industry produced waste materials easily.

6.2.1 Solid chemical waste

Solid chemical wastes are found in solid form in normal temperature. Chemical industries often produce good share of solid waste most of which are metallic waste. These solid wastes get into the environment in many ways, mostly when they are improperly disposed or dumped in the landfill sites. Since most of the solid wastes are metallic waste, they can leach and get mixed with groundwater (Schwab et al., 2007). Presence of metallic contaminants in the groundwater is responsible for extremely negative health consequences; for instance, arsenic contamination leads to arsenicosis, iron contamination leads to anemia, and so on. Apart from the metallic waste, sometimes polymers such as plastic, polyethylene, etc. are also released as solid chemical waste that are one of the major contributors to environmental pollution (Datta and Kopczyńska, 2016). Additionally a good share of packaging and containers are also produced as commonly produced solid waste from chemical industries. Thus, chemical industry can also contribute in solid waste generation.

6.2.2 Liquid chemical waste

Among other different forms of waste released from chemical industries, liquid waste is the most common chemical waste, since most of the industries use the raw materials in liquid form and even many of their products are in liquid form. As a result, liquid waste is the most common class of chemical industry produced waste. As per the raw materials, products and processes involved with the corresponding industries, the liquid chemical waste might be of various types. For instance, the petroleum and refining industries produce significant amount of oil as liquid waste. These oils are split often in the aquatic environment and put a threat to the lives of aquatic species, useful microbes and aquatic plants (Tangchirapat et al., 2007). Additionally, in other industries, oil-like liquid wastes are also very common that are released in the environment getting mixed with wastewater (Caputo et al., 2003). Even highly toxic chemicals are also released as liquid waste from the chemical industries. For instance, the chemical industries that use several carcinogenic or radioactive components, release some of their carcinogenic and radioactive waste with their wastewater. Sometimes, the concentration of these highly toxic materials are so less that there is not much sophisticated technological set up to isolate these (Padhi, 2012). But after being released in the environment, these wastes are bioaccumulated and biomagnified when they enter into the food chain that lead to occurrence of tumors, cancer, and other severe health conditions. Besides, release of radioactive materials in our environment also hampers the lives of other animals, microbes, and plants. Fertilizer industries, another common chemical industry, are also found to release nitrogen containing liquid waste that cause an imbalance of nitrogen content in the environment negatively affecting plant growth and cognitive development of animals (Singh, 2009). Thus, such liquid chemical waste can affect different entities of our environment drastically.

6.2.3 Gaseous chemical waste

Gaseous wastes are of gaseous state in normal temperature. Most of the gaseous wastes produced in the chemical industries are produced as byproducts of the industrial processes. Among the commonly produced gaseous chemical waste, carbon containing gaseous waste is the most common gaseous waste that includes carbon dioxide, carbon monoxide, methane, ethane, and others. The second most common is the nitrogen-containing gaseous waste such as nitrogen dioxide, nitric oxide, nitrous oxides, and others. Along with these simple gaseous substances, there can be other complex gaseous wastes that are released often as the diluted form of large chemicals. As per the type of gaseous waste released, the corresponding hazards can be different. For instance, release of carbon containing gaseous waste is responsible for global warming and climate change. Nitrogen containing gaseous waste is responsible for acid rain, suppress plant growth, cause water pollution, and threaten the lives of different species. In addition to these, several genotoxic and cytotoxic waste are also found as gaseous waste released from the chemical industries which should be converted into other less hazardous gaseous waste before emission (Bauer et al., 1992; Abubackar et al., 2011). Considering different hazards that such waste may cause, these gaseous waste are often filtered using bioreactors either to convert into less hazardous gaseous waste or to minimize the concentration of the hazardous gas emission (Kasperczyk et al., 2012). Because of being in the gaseous form, human health consumes these wastes mostly through inhalation that gradually affects the respiratory system and blood flow.

6.3 Source-level waste minimization initiatives by the chemical industries

Source-level waste minimization initiatives that can be adopted by chemical industries might be quite compatible with many other industries. Like other industries, process modification, recycling, and policy interventions can be regarded as the most suitable source-level waste minimization initiative for chemical industries. Along with these, prerelease waste treatment is also important for waste source reduction in the chemical industries (Fig. 6.1).

6.3.1 Product substitution, process modification, and technological advancement

Like other industries, process modification can be regarded as the most suitable source-level waste minimization initiative for different chemical industries. Process modification do not merely denotes the installing newer and more expensive processes or systems; instead process modification with an aim to waste minimization should focus more on efficiency enhancement, planned regulations, sustainability assurance, and many other features that will prevent waste generation (Stach et al., 2018). Chemical substitution

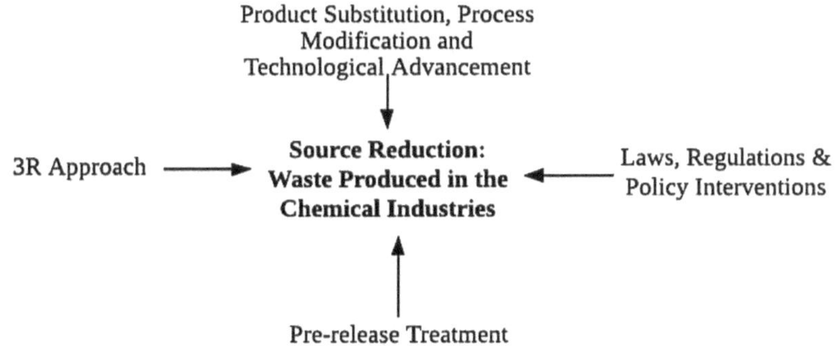

Figure 6.1 ***Overview on suitable source-level waste minimization initiatives by the chemical industries.***

can be regarded as a suitable process modification initiative for chemical industries. Since chemical industries often has several alternative options to be chosen as suitable alternatives, the less waste generating options should be chosen. Aside from this, source-level waste minimization in the chemical industries should also focus to minimize the toxicity of waste. For instance, the chemical industries can adopt green chemistry principles in this regard. Since green chemistry aims to modify industrial chemical processes sustainably without compromising the monetary profit, product substitution mechanism may incorporate green chemistry principles so that their corresponding waste generation is minimized without facing any significant loss (Clark, 2007). Same goes with process modification and technological advancement of the chemical industries. Since the principles of green chemistry is highly focused on sustainable chemical usage modification, while choosing any alternative technology or process or product, in depth chemical alternative assessment should be conducted (Lavoie et al., 2010). Thus, the product options, processes, and technologies should be substituted and modified to minimize source-level waste generation from chemical industries.

6.3.2 3R approach

Implementing 3R approach can also be regarded as a suitable source-level waste minimization initiative for the chemical industries. Primarily, chemical industries should reduce unnecessary chemical usage and thus they should prohibit chemical misuse. Such reduction will not only minimize expenditure spent on excess and unnecessary chemicals but also assure waste minimization caused from excess chemical usage. However, the success of reduction initiative in such a vast industrial sector depends on suitable planning and regulations tremendously (Leist et al., 2008). Second, reuse of chemicals should be prioritized. Especially since a lot of packaging materials and containers are produced in the chemical industries, these should be properly managed so that they can be reused instead of disposing them improperly. Sometimes, in such cases, primary and basic treatments are also needed for sustainable reuse. Even if reuse initiatives will cost some capital cost, the long-term cost savings through resource recovery and through waste minimization will be worth to adopt several approaches to ensure sustainable reuse (Lin and Chen, 1997). Along with reduction and reuse, introducing recycling strategies is also needed for sustainable source-level waste minimization in the chemical industry. Since the major purpose of recycling is to produce valuable resources from waste adopting suitable initiatives, immediate postgeneration chemical waste recycling will assure both resource recovery from efficient mechanism and waste minimization at the same time. Especially in case of chemical waste recycling, some suitable recycling strategies might be resource recovery (i.e., metal extraction, chemical reuse, etc.), wastewater treatment, and so on. Since chemical industries do not produce much of the biodegradable waste, recycling strategies like composting are not appropriate for the specific context.

6.3.3 Prerelease treatment

Chemical industry produced waste should be treated before release to the environment. It is necessary to do so, since such chemical waste can be extremely hazardous for the environmental entities. Such prerelease treatment should incorporate several biochemical treatment initiatives. In this regard, microbial remediation can be regarded as a suitable strategy. There are specific microbes that can reduce certain chemical waste. For example, using microbial bioremediation, certain metallic waste can be reduced. This can be even possible for several complex chemical wastes such as nitro aromatic compound, cyclic organic waste, and so on (Kulkarni and Chaudhari, 2007). Aside from this, physical or chemical processes can also be implemented. Physical processes may include filtration, sedimentation, and so on (Chu et al., 2003). And chemical prerelease treatment processes might be as simple as acid-base reaction to electrolysis and other complex reaction series. Such prerelease treatment is even more important for specialized chemical industries, since in most of the cases, the chemical waste produced by such specialized chemical industries are highly toxic. For example, chemical waste management from semiconductor industry may need to incorporate certain chemical treatment initiatives (Shen et al., 2018). Even chemical wastes generated from pharmaceutical industry often require to be diluted and treated before disposal (Kadam et al., 2016). Along with these, sometimes thermal treatment of chemical waste might also be appropriate. Thus, as per the necessity and demand of the industry, prerelease treatment of the chemical waste might be assured.

6.3.4 Laws, regulations, and policies interventions

Like other industries, laws, regulations, and policies interventions are also needed for chemical industries. First, standardization of disposed chemical waste is a must. All the chemical industries should oblige certain standardization that should be preset by the government or other regulating agencies. In doing so, the industry themselves should adopt certain technologies, initiatives, and approaches. The government can also incentivize the industries that could successfully reduce their sources of chemical waste. Along with these, training and education of chemical waste minimization should be assured for the employees working in different chemical industries so that they can design suitable strategies and contextualize relevant chemical waste minimization initiatives as per their corresponding industrial context. Thus, every role players of the chemical industry should come forward to minimize their waste sources under certain regulations.

References

Abubackar, H. N., Veiga, M. C., & Kennes, C. (2011). Biological conversion of carbon monoxide: rich syngas or waste gases to bioethanol. *Biofuels, Bioproducts and Biorefining*, *5*(1), 93–114.

Allied Market Research (2020). Specialty chemicals market to reach $953.9 billion by 2027 at 5.0% CAGR: AMR. *GlobalNewswire*. Available from: https://www.globenewswire.com/news-release/2020/06/10/2046310/0/en/Specialty-Chemicals-Market-to-Reach-953-9-Billion-by-2027-at-5-0-CAGR-AMR.html.

Arora, A., Fosfuri, A., & Gambardella, A. (2001). Specialized technology suppliers, international spillovers and investment: evidence from the chemical industry. *Journal of Development Economics*, *65*(1), 31–54.

Banerjee, S. B., Iyer, E. S., & Kashyap, R. K. (2003). Corporate environmentalism: Antecedents and influence of industry type. *Journal of Marketing*, *67*(2), 106–122.

Bauer, S., Werner, N., Wolff, I., Damme, B., Oemus, K., & Hoffmann, P. (1992). Toxicological investigations in the semiconductor industry: II. Studies on the subacute inhalation toxicity and genotoxicity of gaseous waste products from the aluminium plasma etching process. *Toxicology and Industrial Health*, *8*(6), 431–444.

BCC Research (2020). *Market research reports: chemicals*. Available from: https://www.bccresearch.com/market-research/chemicals.

Caputo, A. C., Scacchia, F., & Pelagagge, P. M. (2003). Disposal of by-products in olive oil industry: waste-to-energy solutions. *Applied Thermal Engineering*, *23*(2), 197–214.

Chaaban, M. A. (2001). Hazardous waste source reduction in materials and processing technologies. *Journal of Materials Processing Technology*, *119*(1–3), 336–343.

Chu, C. P., Lee, D. J., & Tay, J. H. (2003). Gravitational sedimentation of flocculated waste activated sludge. *Water Research*, *37*(1), 155–163.

Clark, J. H. (2007). Green chemistry for the second generation biorefinery—sustainable chemical manufacturing based on biomass. *Journal of Chemical Technology & Biotechnology: International Research in Process, Environmental & Clean Technology*, *82*(7), 603–609.

Datta, J., & Kopczyńska, P. (2016). From polymer waste to potential main industrial products: Actual state of recycling and recovering. *Critical Reviews in Environmental Science and Technology*, *46*(10), 905–946.

Domazlicky, B. R., & Weber, W. L. (2004). Does environmental protection lead to slower productivity growth in the chemical industry? *Environmental and Resource Economics*, *28*(3), 301–324.

Kadam, A., Patil, S., Patil, S., & Tumkur, A. (2016). Pharmaceutical waste management an overview. *Indian Journal of Pharmacy Practice*, *9*(1), 3.

Kasperczyk, D., Bartelmus, G., & Gąszczak, A. (2012). Removal of styrene from dilute gaseous waste streams using a trickle-bed bioreactor: kinetics, mass transfer and modeling of biodegradation process. *Journal of Chemical Technology & Biotechnology*, *87*(6), 758–763.

Kulkarni, M., & Chaudhari, A. (2007). Microbial remediation of nitro-aromatic compounds: an overview. *Journal of Environmental Management, 85*(2), 496–512.

Lavoie, E.T., Heine, L.G., Holder, H., Rossi, M.S., Lee, R.E., Connor, E.A., et al. (2010). Chemical alternatives assessment: enabling substitution to safer chemicals, Environmental Science & Technology, *44*(24), 9244–9249.

Leist, M., Kadereit, S., & Schildknecht, S. (2008). Food for thought on the real success of 3R approaches. *ALTEX-Alternatives to Animal Experimentation, 25*(1), 17–32.

Lin, S. H., & Chen, M. L. (1997). Treatment of textile wastewater by chemical methods for reuse. *Water Research, 31*(4), 868–876.

Padhi, B. S. (2012). Pollution due to synthetic dyes toxicity & carcinogenicity studies and remediation. *International Journal of Environmental Sciences, 3*(3), 940–955.

Schwab, P., Zhu, D., & Banks, M. K. (2007). Heavy metal leaching from mine tailings as affected by organic amendments. *Bioresource Technology, 98*(15), 2935–2941.

Shen, C.W., Tran, P. P., & Minh Ly, P.T. (2018). Chemical waste management in the US semiconductor industry. *Sustainability, 10*(5), 1545.

Singh, B. (2009). Treatment of spent catalyst from the nitrogenous fertilizer industry—A review of the available methods of regeneration, recovery and disposal. *Journal of Hazardous Materials, 167*(1–3), 24–37.

Soni, M. (2020). Top 6 types of chemical industries: pollution. *Environmental Pollution.* Available from: https://www.environmentalpollution.in/industrial-pollution/top-6-types-of-chemical-industries-pollution/2259.

Stach, T., Dräger, J., & Huy, P. H. (2018). Nucleophilic substitutions of alcohols in high levels of catalytic efficiency. *Organic Letters, 20*(10), 2980–2983.

Tangchirapat, W., Saeting, T., Jaturapitakkul, C., Kiattikomol, K., & Siripanichgorn, A. (2007). Use of waste ash from palm oil industry in concrete. *Waste Management, 27*(1), 81–88.

CHAPTER 7

Source reduction and waste minimization in food industry

7.1 Introduction

Food industry is one of the most important industrial settings for any country in terms of the huge public health and economic concerns associated with this industry. Unlike many other industries, food industry is a highly decentralized and less coordinated industrial system due to the highly segregated roles of different stakeholders under the industrial setting. Food industry denotes a complex global industry that clusters many small- to large-scale industries or business that are related with food item production, processing, supply, and distribution (Bigliardi and Galati, 2013). In broader terms, food industry also incorporates food preservation, conversion, packaging, and many more complex food-related processes. The industry itself is very diverse, since different sectors of food industries deal with different food items, different food processing mechanisms, and different stages of food supply and management. In addition to these, the diversity of the industrial role players, in terms of their socio-economic status, geographic location and others, is another factor that makes this industry a highly segregated industrial setting. However, even if the whole industrial set up of food industry seems quite complex, it is a highly organized industry in most of the countries, since disruption in this industry may lead to sudden breakdown of the public health and economic scheme of any country.

Food being the basic need of human, makes food industry more important industry than any other industries. Understanding the unavoidable dependency on a sustainable food industrial system for human welfare, it is necessary to regulate, maintain, and improvise the overall food market situation carefully. Additionally, the economic contribution of food industry is also very important for global economy. The CAGR of smaller segments of food industry are estimated to experience a significant rise in near future; for example, global organic food market is estimated to experience a compound annual growth rate (CAGR) of 11.5% from 2019 to 2024, protein food market will experience a CAGR of 9.6% in between 2018 and 2025, and so on (BCC Research, 2019). Most importantly global market of

Source Reduction and Waste Minimization
http://dx.doi.org/10.1016/B978-0-12-824320-6.00007-1

packaged or frozen food items will experience a huge rise, since people are trying to minimize their food processing time in this modern era to save up their time for other purposes (GlobeNewswire, 2020). Therefore, food items that require long preparation or processing time will have a comparatively lower CAGR. Thus, as per the necessity and choice of the people, the compound annual growth rate of different components of food industry varies significantly. Along with these, the trend of food industry in a country often determines several health factors of the country people. For example, if a country fails to promote some fortified food items in their food market, there will be negative public health consequences as a result of micronutrient deficiency among the citizens.

However, realizing all these sensitive aspects of food industrial set-up, food waste minimization initiatives should also be handled very carefully so that a proper balance of both the economy and health is maintained. The estimated amount of annual food waste generation around the globe is around 1.3 billion tons that is equivalent to one-third of the total produced food items for human consumption (Reset, 2020). This denotes that only a net total of two-third of the produced food products are consumed. This also indicates the large amount of misused food items. Understanding the consequences that this vast amount of food waste generation may bring about, it is necessary to take immediate initiatives to adopt source-level food waste minimization. Many of such initiatives have already been introduced. The global food waste management initiatives reached a market worth of 31.3 billion USD in 2019 that is predicted to reach up to 43.1 billion USD by 2024 with a CAGR of 6.6% in between 2019 and 2024. Unfortunately, in many cases, these initiatives are not adopted at the source-level for which pregeneration waste prevention initiatives are not quite common in case of food industry. As a result, competing with the ever rising global population growth, the amount of food waste generation is rising every year. This is a huge misuse of valuable resources and can also increase the environmental burden of waste. If source-level minimization of food waste is assured, such vast amount of waste generation can be easily prohibited by putting an end to food item misuse and by producing useful resources from the food waste. This denotes the urgency to take source-level food waste minimization initiatives.

However, in this chapter, some of the commonly produced food waste type has been discussed along with their corresponding hazards. And it also explores some potential source-level food waste minimization strategies that can be implemented for food waste minimization.

7.2 Commonly produced waste in the food industries: source and biogeochemical hazards

As mentioned earlier, food industry is a significantly segregated industry with numerous number of small to large role players. The total amount of produced food waste around the world is so vast that if source-level pregeneration waste minimization initiatives or immediate postgeneration initiatives are not taken, the total amount of produced food waste will keep rising uncontrollably. Teuber and Jensen (2016) defines food waste as any food components or inedible parts of the food, removed from the food supply chain which can be either recovered or disposed. Besides, any food item that could have been eaten, but somehow, if it is excluded from the human food supply chain or from any point of the supply chain, then it can also be regarded as a component of food waste (Teuber and Jensen, 2016). Since most of the food wastes are generated directly from the food items, usually, food wastes are considered to be less hazardous comparing with other industrial waste classes. The extent and trend of food waste generation is not similar in different contexts. Geographic parameters, availability of food and many other variables influence the types and amount of generated food waste. Especially the economic state of the stakeholders affect the amount and types of generated food waste significantly; for instance, the amount and type of food waste generated in the low income households are significantly different from the high income households (Porpino et al., 2015). However, since food industry is a significantly large waste generating industry associated with a lot of variables, the produced waste in this industry can be classified under several criteria. Table 7.1 classifies some of the major classes of food waste generated in the food industry along with the corresponding biogeochemical hazards that might be caused due to such waste.

7.3 Source-level waste minimization initiatives by the food industries

Food wastes are one of the commonly produced waste classes by all the individuals in every single day. Though the toxicity and diversity of food industry produced waste are not as high as many other industries, considering the large amount of generated food waste, source level waste minimization initiatives should be introduced at large scale so that all the role players can understand their roles and benefits out of this source reduction initiative adoption. Though there is a common trend of assigning the responsibilities of food waste minimization to the municipal waste management authorities,

Table 7.1 Commonly produced waste in the food industries along with their corresponding hazards (Teuber and Jensen, 2016; Sánchez-Rodríguez et al., 2017; Farr-Wharton et al., 2014; Evans, 2012; Aschemann-Witzel et al., 2015).

Types	Sources and features	Hazards
Organic crop residue	These wastes are mostly agricultural wastes that are produced in agricultural fields. Since such crops are components of food items, the organic crop residue can be considered as food waste. These crop residues are the materials which are usually left in the agricultural field right after the crop is harvested.	• Cannot be regarded as highly hazardous or toxic waste. • Often makes the surrounds dirty. • Hampers the lives of aquatic species when these get mixed in the aquatic environment. • Carry dirt and dusts.
Catering waste	Catering wastes are those wastes that are produced during the cooking or other food processing initiatives. Usually such wastes are generated in the kitchens, restaurants, fast food shops, and during other food processing steps. The inedible or wasted food items produced during the cooking process are known as catering waste. Catering waste can be of various types, as per the available food items during cooking. Catering waste may include cooking oil, meat, fish, bakery, flour, rice, and many other food items. In most of the cases, these wastes are produced as a result of food loss or food misuse. Besides, sometimes, catering wastes are produced as by-product of cooking processes that cannot be avoided to be generated.	• Though such wastes do not usually contain any toxic components, many of the catering waste are found to get toxic when released in the environment due to their environmental interaction with other substances. • Catering waste can kill aquatic species when getting mixed with the aquatic environment. • Exposure to old catering waste may initiate vomiting, dizziness, and weakness due to the bad odor caused from the decomposition of catering waste. • Oil-based catering waste often kills microorganisms or hampers plant growth negatively affecting their nutrient intake. • Organic catering waste is found to cause odor pollution when disposed in the open space without proper regulations.

Packaging	Packaging is one of the largest food waste generating components. Though many other industries are found to utilize good share of packaging products, it is a more significant issue in case of food industry, since food industry is found to use more amount of packaging due to safety maintenance and proper food distribution mechanism. Packaging can be regarded as one of the most hazardous food waste, since most of the packaging products are not biodegradable and these wastes often clog, cluster, or accumulate here and there putting a threat to the surrounds. Packaging used in the food industries should be properly managed and disposed to minimize its corresponding hazards.	• Packaging products are often made of polyethylene, plastics, nylon, and other polymers which remain in the environment for a long period, since they are not easily perishable. • When packaging products are released in the aquatic environment, they prohibit the free movement of aquatic animals putting a threat to their existence. • Long-term accumulation of packaging products cause soil pollution initiating ecological imbalance and other environmental hazards. • Since most of the packaging wastes are part of solid waste, their treatment and management become tougher when they are released in the environment randomly.
Domestic Waste	Domestic food wastes are those food wastes that are produced at a household level. Since households are very small entities under the decentralized and highly segregated roof of food industry, it is necessary to understand the role of domestic food waste for better management or minimization of these food waste produced at a household level. Domestic food waste includes a variety of food waste that includes expired food waste, packaging generated from food items stored in the households, by-products produced during food processing, supply or distribution and many more different types of waste. Domestic food waste may also incorporate the wasted food produced by the households.	• Domestic food waste may cause several different types of hazards in the environment and for human health. Since domestic food wastes are generated at a household level, they can be easily exposed by the household members and cause biochemical component-borne health issues. Besides, when these domestic food wastes are released in the environment, they gradually cause soil, water, air, and odor pollution. • Since most of the domestic food waste are biodegradable waste, their health and environmental consequences are not that deadly.

(*Continued*)

Table 7.1 Commonly produced waste in the food industries along with their corresponding hazards (Teuber and Jensen, 2016; Sánchez-Rodríguez et al., 2017; Farr-Wharton et al., 2014; Evans, 2012; Aschemann-Witzel et al., 2015). (*Cont.*)

Types	Sources and features	Hazards
Animal by-products	Animal by-products are one of the most hazardous classes of food waste produced during different stages of food production, supply and distribution. Such wastes usually include bloods/tissues/residues produced during slaughtering of animals, meat, fish, expired non-vegetable food items, and others.	• Animal by-products may lead to infectious or pathological waste transmission. • Expired meats, fishes, and other food items may cause severe gastrointestinal problems and other health issues. • Environmental exposure to such waste kills useful microbes, hampers plant growth, and degrades the environmental parameters.
Mixed-food waste	When all the earlier-mentioned food wastes are released together in mixed form, they are known as mixed food waste. Mixed food wastes are comparatively more hazardous than individual classes of food waste, since in case of mixed food waste, the substances present in the food waste interact and react with each other generating more toxic substances. Mixed food waste can be generated by any of the entities of the food industry. Besides, such mixed food waste are often result of food misuse, improper management, lack of waste segregation initiatives, and many more improper waste management initiatives.	• Mixed food wastes are often dumped into the landfill sites thus contributes to GHG emissions after decomposition and promotes global warming and climate change. • Exposure of mixed food waste in the environment degrades the environmental parameters for air, water, soil, and other components. • Mixed food waste is responsible for degraded respiratory and gastrointestinal condition since these promote emission of harmful gaseous and liquid substances.

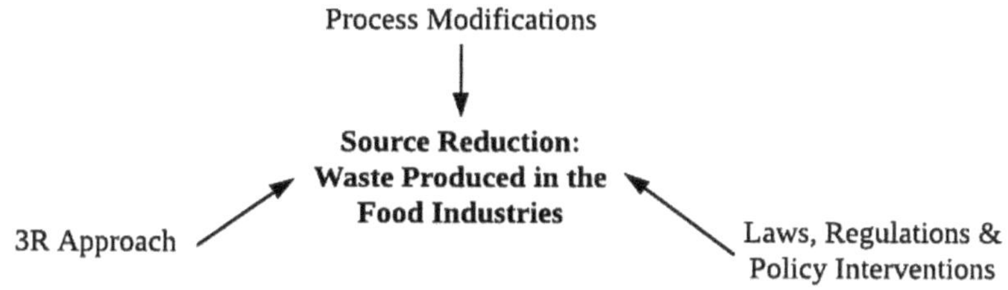

Figure 7.1 ***Overview on the source-level waste minimization initiatives for food industries.***

unless subsequent amount of initiatives are undertaken by the industrial authorities, the every rising amount of waste cannot be tackled only through the initiatives of the supporting agencies. (Abdoli and Azimi, 2010). The broader classes of source-level waste minimization initiatives that can be undertaken by the food industries might be categorized under three major classes mentioned in Fig. 7.1.

7.3.1 Process modifications

Process modification is one the most common waste source reduction strategies; especially in case of industrial waste generation, small- to large-scale modification of the industrial processes associated with the supply chain, may lead to sharp reduction in the waste generation. For food industry, this is even more appropriate since food waste generation can be reduced, if the associated processes, technologies, and products are efficient enough. Since a significant share of the food waste produced by different entities of the food industry is caused due to the misuse or loss of food items, only efficiency increase of the entities of food industries may reduce such misuse-led food waste generation drastically (Henz and Porpino, 2017). Food loss or increased share of edible food items as food waste, can be regarded as a major flaw of the supply chain and operation mechanism of food industry. Only efficient process modification and willingness of the industrial owners can help to minimize significant share of food loss induced food waste generation. Betz et al. (2015) argue that one of the largest food waste generating Swiss industries estimated 18% loss of the food input among which the percentage of the avoidable losses were 13.5%. This shows how efficient food loss or misuse prevention mechanism can help to minimize food waste generation drastically. Additionally, sometimes even if some waste management initiatives are introduced by industrial authorities, these are not efficient enough in many cases (Ye et al., 2018). As a result, the benefits cannot be achieved as expected. Efficiency of such processes should also be increased for sustainable source-level food waste minimization.

Along with process modification through efficiency enhancement, more effort should be put toward efficient process planning initiatives. Since food industry is not a much aggregated industry, introducing efficient planning initiatives are not often possible for such highly segregated industrial context. In this regard, the industrialists themselves should play the role of efficient planning for their corresponding industrial setting (Soman et al., 2007). Unless they upgrade the plans at a regular basis and prefer not to stick with the conventional initiatives undertaken by former practices, the ever-rising food waste generating trend cannot be decelerated. For example, nutritional programs can be introduced to prevent misuse or loss of food products causing generation of higher amount of food waste. Launching new nutritional program in schools contextualizing the needs and conditions of the students has been found to bring about significant reduction in the daily generated amount of food waste (Byker et al., 2014). Likewise, every role players associated with different stages of the food industry should come forward from their respective places so that the overall food waste generation can be tackled.

7.3.2 3R approach

Like other industries, 3R approach can also be a suitable source-level waste minimization strategy for food industry. Primarily reduction and reuse can be introduced that will ensure drastic minimization of the total amount of produced food waste. For example, to implement reduction strategy, misuse of food items or loss of food items should be prohibited. Therefore, reduction, being the first step of 3R approach might be adopted in the food industry in different ways. First, reduction approach should be aimed to minimize misuse of food, especially in the developed countries where food resources are already adequate. This practice of preventing food misuse cannot be forced upon the individuals, rather the educational system should incorporate teaching the consequences of food waste misuse and also teach the students about their role in terms of minimizing their personally produced food waste. Thus, the students will grow the habit of food misuse prevention from an early age. Training facilities for employees working in the food industrial sectors should also be introduced so that they can adopt waste prevention initiatives from their respective sectors. Additionally, the restaurants, shops, factories, and other entities of the food industry should be imposed taxation on food waste generation and landfilling tax on dumping food waste in the landfill sites so that they are also concerned to adopt several process modification initiatives for source level food waste minimization. Along with these, using unnecessary packaging materials made of nonbiodegradable matters should also be reduced through industrial initiatives and government's regulations.

In addition to waste reduction, waste source reduction in the food industries should also introduce subsequent amount of reuse initiatives. Reuse denotes utilizing the produced waste as resources for other initiatives. Most of the food waste being biodegradable waste might be reused for several other purposes. For example, some of the catering waste produced in the food industries can be directly utilized as fertilizing nutrients for plant growth. This may include vegetable waste, used tea leaves, expired cooked food items, and others. Agricultural food waste also has significant reuse potentials. For example, crop residues are often clustered in the river embankments as a flood prevention strategy or to strengthen the embankments. Besides, packaging materials can also be reused several times. If proper waste collection and segregation mechanisms are ensured for food waste industry produced packaging waste sorting, reusing these wastes for other purposes will be easier. Such initiative has been already introduced at several different food industries. For instance, closed loop packaging system can be introduced where the packaging can be used for multiple times (Accorsi et al., 2020). Such closed loop waste minimization initiative can be introduced by business initiatives or sometimes even by the food industry themselves. However, even if business initiatives or government initiatives are taken in this regard, unless the food industries take responsibilities in reusing their packaging waste, the reuse mechanism won't be efficiently implemented, since the food industries have the maximum potentials to reuse and repackage newer food items using previously generated packaging food waste. Thus, the industrial responsibilities for reusing the produced food waste should be enhanced.

Recycling, being the final stage of 3R approach, can be introduced for food waste minimization. Composting is a very common recycling initiative that can be considered as a suitable immediate postgeneration or nearly source-level waste minimization initiative. As mentioned in the previous chapters, composting is a very well-known decomposition strategy where organic wastes are put under certain chemical and physical condition to be decomposed in presence of oxygen. Since most of the food waste contain good amount of carbohydrate, proteins, and fats, these wastes can be regarded as very efficient raw materials for composting. In this regard, suitable microbes and conditions should be chosen as per the materials present in the food waste to assure maximum efficiency (Chang et al., 2006). Even sometimes it is necessary to constantly change the imposed conditions to determine the most efficient composting condition for which multiple experiments should be conducted (Li et al., 2013). Thus, this recycling approach, composting, will produce useful humus or soil nutrients that will help to promote plant growth. This can also be regarded as a resource recovery system, since it recovers economically valuable resources from waste. Along with this, other recycling initiatives should also be introduced focusing recycling of food waste so that valuable resources can be recovered from the produced food waste.

7.3.3 Laws, regulations, and policies interventions

Unless the government takes the leading role to encourage and influence the industrial bodies to adopt suitable source-level waste minimization initiatives, the industrial authorities cannot be forced to adopt such initiatives. Therefore, the intervention of the government is also important for such a highly segregated industry like food industry where every individual's responsibilities are important. Like other industries, government or other regulating agencies may impose standardization limits to the industrial authorities that they are bound to maintain before releasing the waste in the environment. For instance, the slaughter houses can be asked to deeply clean the bloods and animal residues using certain chemical disinfectants. The liquid food waste producing industries can be asked to introduce primary level chemical or biological treatment before disposing their liquid waste in the rivers or canals. The standardization initiatives should also be strictly monitored and evaluated over time to assure that the industries are taking sincere initiatives to follow the standards. Additionally, some other types of commonly followed regulations should be maintained for sustainable waste prevention for food industries. This might include providing incentives or subsidies to the food industries to set up suitable source-level waste minimization initiatives. The concept of circular economy can be followed to assure maximum resource efficiency through proper circulation of resources in the market. Additionally, ban can also be imposed for certain items associated with the food industries. For example, since many of the food item packaging options are often considered as unsustainable, some of the most hazardous packaging options can be banned by the regulating authorities providing suggestions on suitable alternatives. Thus, the government's regulations and interventions will promote source-level waste minimization in the food industries.

References

Abdoli, M.A., & Azimi, E. (2010). Municipal waste reduction potential and related strategies in Tehran. *International Journal of Environmental Research, 4*(4), 901–912.

Accorsi, R., Baruffaldi, G., & Manzini, R. (2020). A closed-loop packaging network design model to foster infinitely reusable and recyclable containers in food industry. *Sustainable Production and Consumption, 24*, 48–61.

Aschemann-Witzel, J., De Hooge, I., Amani, P., Bech-Larsen, T., & Oostindjer, M. (2015). Consumer-related food waste: Causes and potential for action. *Sustainability*, 7(6), 6457–6477.

BCC Research (2019). *Market research reports: food*. Available from: https://www.bccresearch.com/market-research/food-and-beverage.

Betz, A., Buchli, J., Göbel, C., & Müller, C. (2015). Food waste in the Swiss food service industry–Magnitude and potential for reduction. *Waste Management, 35*, 218–226.

Bigliardi, B., & Galati, F. (2013). Innovation trends in the food industry: the case of functional foods. *Trends in Food Science & Technology, 31*(2), 118–129.

Byker, C. J., Farris, A. R., Marcenelle, M., Davis, G. C., & Serrano, E. L. (2014). Food waste in a school nutrition program after implementation of new lunch program guidelines. *Journal of Nutrition Education and Behavior, 46*(5), 406–411.

Chang, J. I., Tsai, J. J., & Wu, K. H. (2006). Thermophilic composting of food waste. *Bioresource Technology, 97*(1), 116–122.

Evans, D. (2012). Beyond the throwaway society: Ordinary domestic practice and a sociological approach to household food waste. *Sociology, 46*(1), 41–56.

Farr-Wharton, G., Foth, M., & Choi, J. H. J. (2014). Identifying factors that promote consumer behaviours causing expired domestic food waste. *Journal of Consumer Behaviour, 13*(6), 393–402.

GlobeNewswire (2020). *Global packaged food market is expected to grow with a healthy CAGR over the forecast period from 2020-2026*. Available from: https://www.globenewswire.com/news-release/2020/08/28/2085717/0/en/Global-packaged-food-market-is-expected-to-grow-with-a-healthy-CAGR-over-the-forecast-period-from-2020-2026.html.

Henz, G. P., & Porpino, G. (2017). Food losses and waste: how Brazil is facing this global challenge? *Horticultura Brasileira, 35*(4), 472–482.

Li, Z., Lu, H., Ren, L., & He, L. (2013). Experimental and modeling approaches for food waste composting: A review. *Chemosphere, 93*(7), 1247–1257.

Porpino, G., Parente, J., & Wansink, B. (2015). Food waste paradox: antecedents of food disposal in low income households. *International Journal of Consumer Studies, 39*(6), 619–629.

Reset (2020). Global food waste and its environmental impact. *Reset: Digital for Good*. Available from: https://en.reset.org/knowledge/global-food-waste-and-its-environmental-impact-09122018.

Sánchez-Rodríguez, A. R., Hill, P. W., Chadwick, D. R., & Jones, D. L. (2017). Crop residues exacerbate the negative effects of extreme flooding on soil quality. *Biology and Fertility of Soils, 53*(7), 751–765.

Soman, C. A., van Donk, D. P., & Gaalman, G. J. (2007). Capacitated planning and scheduling for combined make-to-order and make-to-stock production in the food industry: An illustrative case study. *International Journal of Production Economics, 108*(1–2), 191–199.

Teuber, R., & Jensen, J.D. (2016). *Food losses and food waste: extent, underlying drivers and impact assessment of prevention approaches*. Department of Food and Resource Economics, University of Copenhagen. IFRO report No. 254.

Ye, M., Liu, J., Ma, C., Li, Y. Y., Zou, L., Qian, G., et al. (2018). Improving the stability and efficiency of anaerobic digestion of food waste using additives: a critical review. *Journal of Cleaner Production, 192*, 316–326.

CHAPTER 8

Source reduction and waste minimization in energy production industry

8.1 Introduction

Energy producing industry is an integral sector for modern era. Energy producing industries incorporate the production, sales, supply, and distribution of energy among the stakeholders. Here, the energy producing industries can be of various types that may include fuel, electricity, oil/gas drilling industries, and others. However, in a broader sense, nowadays, energy industry can be classified under non-renewable and renewable energy. Non-renewable energy production industry includes oil, petroleum produces, fossil fuel, gasoline, diesel, nuclear, and other energy sources whereas renewable energy sources include hydropower, wind power, solar energy, biofuels (i.e., ethanol) and others. Energy is a basic human need nowadays since all different forms of energy are always needed in our everyday life. Energy production industries use a variety of raw materials that determines if the energy source is renewable or not. Apart from this, the purpose of the produced energy can also be regarded as a factor that may also classify energy producing industry under several different classes. For instance, energy producing industry might be classified as cooking fuel, transportation fuel, electricity supply, and so on. Whatever classification strategy is followed, usually processes, materials, and strategies of most of the energy producing industries are more or less similar. This vast energy market is highly relatable with the socio-economic and technological growth of the country.

Energy market is quite vast and its economic growth may vary among different sectors of the energy industry. For instance, co-current urge to shift from conventional to renewable energy market of energy industry makes the estimated CAGR of renewable energy market higher than the conventional fossil fuel based energy market. Only heat and power generation industry is estimated to experience a CAGR of 6.54% in between 2017 and 2022 making a shift from 1.67 to 2.29 billion of USD worth of industrial

Source Reduction and Waste Minimization
http://dx.doi.org/10.1016/B978-0-12-824320-6.00008-3

investment in this time period. The renewable energy market is estimated to experience around 5% of CAGR by 2025 increasing the market investment up to 2.15 billion of USD by 2025 (Institute of Energy for South-East Europe, 2020). Even this percentage of CAGR is quite high for an emerging industry like renewable energy industry. Besides, as a service market, the growth of energy industry is even higher since it is estimated that energy service market will reach up to the worth of 172.9 billion of USD by 2027 with a CAGR of 14.6% (Grand View Research Incorporated, 2020). All these data show the importance of this specific industry for the economic growth of the country.

However, despite the significant importance of the issue in current context, a lot of challenges are associated with different aspects of the energy industry. Most of the energy industries emit huge amount of toxic chemicals that require a sharp shift from fossil fuels to other sustainable or renewable resources. Though fossil fuel is regarded as the most common energy source, it is not a sustainable energy source for multiple reasons such as, high GHG emission potentials, limited amount of left fossil fuels, unsustainable industrial set up, and many more. This is why, in current world, the urge of shifting from fossil fuels to other renewable energy sources are being prioritized. Renewable energy sources-based energy production offers many a lot of facilities over non-renewable energy resources that includes unlimited availability, environmental friendly, sustainability, and others. Besides, several ongoing innovative initiatives in renewable sector is also indicating an optimistic scenario for sustainable energy future. All these different factors indicate the necessity to substitute fossil fuels by renewables for a sustainable future with adequate amount of energy generation potentials (Leonard et al., 2020). Along with this, the energy market requires to install very advanced technological set up for efficient energy generation mechanism that many of the developing countries are struggling with. The industrialists involved with this industry are often not concerned enough to install sustainable energy production set up causing generation of huge amount of waste by the energy production industry. Thus different challenges make the operation, maintenance, and supply chain situation associated with waste generation even more complex.

This chapter explores some of the commonly produced waste by the energy production industry along with their corresponding biochemical hazards. Additionally, it identifies some of the potential source level waste minimization initiatives for energy production industries.

8.2 Commonly produced waste in the energy production industries: source and biogeochemical hazards

As discussed earlier, energy producing industry is one of the most important industrial sector that aims to meet the people's energy demand. In this process of energy production, several different types of wastes are produced that are commonly regarded as energy generation mediated waste. These wastes are often responsible for occurrence of different hazards and issues. Classifying energy industry produced waste is quite complex since the variety of generated waste in this industry is so huge and diverse in nature that broader categories cannot often be identified. Therefore, Table 8.1 summarized some of the feature and corresponding hazards of the common energy industry produced waste. However, the total amount of produced waste by the energy production industry is not limited to the major waste classes mentioned in the Table 8.1, instead, these are many more different waste classes produced by energy producing industries. That may include, plastics, paper, cardboards, woods, and so on.

8.3 Source-level waste minimization initiatives by the energy producing industries

As mentioned in the Table 8.1, energy producing industries generate many types of highly hazardous waste, it is important to prevent such generation of waste as much as possible for both short-term welfare and long-term sustainability. Like all other industrial waste source reduction mechanisms mentioned in all the previous chapters, source-level waste minimization should be prioritized for even energy producing industries, since it will help to reduced exposure duration and minimize contact with such hazardous waste. Fig. 8.1 summarizes major source-level waste minimization initiatives that might be undertaken by the energy producing industries.

8.3.1 Product substitution

Product substitution can be regarded as a primary and compulsory source-level waste minimization strategy for energy producing industries. In this regard, the major initiative might be fossil fuel-based energy generation mechanism substitution. As mentioned earlier in this chapter, the urgency of fossil fuel substitution might be validated by several corresponding factors such as limited availability of fossil fuels, complex processing strategies, unsustainability, carbon emission potentials, and other. However, only environmental welfare is not the only concern for substituting fossil fuels

Table 8.1 Commonly produced waste in the energy producing industries (Ritchie and Roser, 2018; Ahmed et al., 2011; Steinfeld and Meier, 2004; Ambashta and Sillanpää, 2012; Xu et al., 2017; Ike et al., 2018; Themelis, 2003).

Waste class	Sources and feature	Biochemical hazards
Organic pollutants	Fossil fuel is one of the most common raw materials for energy sector. More than 80% of the world's total energy consumption comes from fossil fuels causing the industry to be one of the leading emitter of fossil-fuel generated waste. Such fossil fuels include, petroleum diesel, coal, tars, and so on. These fuel themselves are huge source of carbon and cause generation of several organic pollutants during the energy generation process and sometimes even after the postgeneration of waste.	• Such organic pollutants are usually released to the environment getting mixed with liquid waste. And when these liquid wastes are released to the environment, soil, and water pollution occurs. • Aquatic parameters degrade significantly causing death of several aquatic species.
Metallic waste	A good share of energy producing industry generated waste is metallic waste that contains good amount of heavy metals and other comparatively less harmful metals. These metallic waste are released either in solid or liquid form. The infrastructural components of the energy industry are usually the metallic waste sources. Additionally metallurgical industries concerning with energy generation also produce good share of metallic waste during higher energy generation schemes. Solid fuel based industries are also often responsible for metallic waste generation as by-products of their energy generation processes. Even renewable energy producing industries such as hydropower or solar energy producing systems also lead to huge amount of metallic waste production after end-use of the technologies.	• Metallic waste with high metallic concentration often leads to several metal-borne health concerns; for example, some common metallic waste borne health hazards might be kidney/liver/lungs dysfunction or damage. • Long-term exposure to metallic waste disrupts cellular transportation mechanism causing cancer or tumors in the long run. • Environmental exposure to metallic waste negatively affects the environmental biodiversity causing environmental degradation.

Air pollutants	Air pollutants emission is one of the major waste issue related concern of the energy producing industries. Energy industry is one of the major air pollutants emitting industry. Due to fossil fuel based energy production, several different air pollutants such as carbon dioxide, carbon monoxide, methane, nitrogen dioxide, nitric oxide, nitrous oxides are released.	• Carbon containing air pollutants cause global warming and climate change in both short- and long-term. • Nitrogen containing air pollutants leads to acid rain, hampered plant growth, and other toxic effects for both terrestrial and aquatic environment.
Radioactive waste	Radioactive waste is one of the leading wastes produced by nuclear energy producing industries. Radioactive materials are usually used for nuclear energy generation that might be found in the waste stream of the nuclear power plants. Such radioactive wastes often contain uranium, radium, tellurium, and other radioactive materials.	• Radioactive wastes are responsible for occurrence of cancer, tumor, and other deadly health concerns. • Impaired growth and hampered physiological conditions. • Unless proper maintenance of radioactive material containing is assured, the exposed plants, animals, and microorganisms will be affected tremendously.

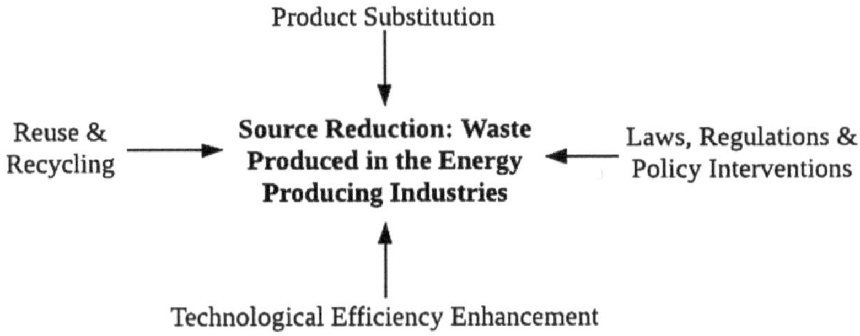

Figure 8.1 ***Major source-level waste minimization initiatives for energy producing industries.***

by other energy sources. Long-term sustainability won't be assured, if such product substitution mechanism is not adopted as early as possible, since many researchers have already predicted that the available stock of fossil fuels to be used for energy production purpose will run out anytime in between 2050 and 2060 (Howarth, 2019). Unless any alternatives are not initiated as early as possible, there will be no scope to cause a sudden shift from fossil fuel to renewables or any other sustainable alternatives (Gielen et al., 2019). Along with this urgent large shift of energy resource substitution, smaller scale substitution is also required. That may include substituting smaller options in the whole supply chain as well. For example, product substitution in terms of the infrastructure construction might be prioritized, since many of the energy producing industries are often found to use age-old material constructed infrastructures that incorporate high metallic contents. There are already good amount of suitable and durable alternative options for more sustainable infrastructure construction (Lin and Chen, 2019). Thus, as per the industrial context, sustainable product substitution mechanism should be assured.

8.3.2 Technological efficiency enhancement

Along with product substitution, suitable technological efficiency enhancement should also be prioritized in this regard. Industrial process inefficiency can be regarded, as one of the leading factors causing large share of waste generation. Especially, in case of energy sector, inefficiency can be regarded as one of the leading cause of large share of waste generation. Efficient energy production and consumption may lead to over 11% growth within next 25 years that denotes the economic urgency and necessity of industrial initiatives for waste source reduction in the energy producing industries (The

National Academics o Sciences Engineering and Medicine, 2020). However, in case of energy industry, technological initiatives might be initiated for different levels; from generation to consumption. Only industrial initiatives to promote technological efficiency won't be of any use, unless the consumers also take the responsibilities to play their role of efficient energy usage from the consumers' side. Energy production and consumption associated technological efficiency should be assured through awareness raise, community mobilization, and through inclusion of rational consumer's behaviors in the educational curriculum. Most importantly, technological assessment should be conducted in this regard, so that the efficiency index of several technologies can be compared determining the most efficient technological innovations (Miao et al., 2018). Thus, all these factors should be deeply assessed during efficient technology assessment.

8.3.3 Recycling and reusing

Recycling and reusing initiatives can also be applicable for energy producing industries like all other industries. Recycling and reusing initiatives might be applicable in this regard in various ways such as, waste to energy generation, resource recovery, pollutant isolation, and in many other ways. The waste to energy concept might be very relatable for this specific industrial context. Technologies are being innovated that captures waste with suitable energy producing potentials. This is a part of waste to energy concept. The energy producing waste should take this into account since waste to energy concept implementation doesn't cause much for energy producing resource purchase and at the same time it also leads to industrial sustainability for long term (Zhang et al., 2017). Second, even if renewable energy is adopted, some share of waste is generated that might be highly hazardous if not treated and managed properly. Solar energy is regarded as one of the most efficient and sustainable source of energy in current context. Unfortunately, though dependency on solar energy leads to reduction of fossil-fuel based GHG emission, currently, end-use products of solar panels are a matter of great concern, since these dysfunctional solar panels cause occurrence of metal leaching, organic pollutant emission, and many other troubles when disposed openly after end use (Bachrach and Blonigan, 2010). Considering this issue, dysfunctional solar panel recycling might be regarded as an urgent technological necessity currently. Though many innovative initiatives are being undertaken in this regard, unfortunately large-scale initiatives of those innovations have not been possible yet (Shin et al., 2017). Sometimes, it is not possible to reuse the dysfunctional solar panels after recycling; in such cases, these panels might be sent to respective authorities who isolate valuable metals and resources out of the dysfunctional solar panels. In this regard, many business and entrepreneurships may grow that are based on silicon isolation, metal recovery from used and dysfunctional solar panels (Lin et al., 2012). Thus, considering different contexts of the energy producing industry, reuse and recycling initiatives might be adopted.

8.3.4 Laws, regulations, and policies interventions

Like other industries, strict laws, regulations, and policies should be adopted by the government to ensure sustainable source-level industrial waste minimization. Like all other industries, laws, regulations, and policies should be adopted in the energy industries so that the corresponding hazards are minimized. As mentioned in all the previous chapters, standardization, incentives, monitoring and evaluation, quality assurance, all these different policies should also be incorporated for source-level waste minimization in the energy producing industries. However, among the other types of regulations, renewable energy promotion should be prioritized. In this regard, the government initiatives for countries such as Germany and Bhutan can be considered as very appropriate example, since these countries could promote successful renewable energy generation scheme, gradually reducing most of their fossil fuel based energy generation schemes. Thus, these countries along with many other countries are targeting to 100% renewable energy generation within first half of this century (Lehr et al., 2008; Young et al., 2007). Thus, promotion of renewable energy, being a potential source level waste minimization initiative might be promoted by the government and other regulating agencies.

References

Ambashta, R. D., & Sillanpää, M. E. (2012). Membrane purification in radioactive waste management: a short review. *Journal of Environmental Radioactivity*, *105*, 76–84.

Ahmed, M.M., Kutty, S.R.M., Shariff, A.M., & Khamidi, M.F. (2011). Petrol fuel station safety and risk assessment framework. In *2011 National Postgraduate Conference* (pp. 1–8). IEEE.

Bachrach, R., & Blonigan, W.T. (2010). *U.S. Patent No. 7,759,158*. Washington, DC: U.S. Patent and Trademark Office.

Gielen, D., Boshell, F., Saygin, D., Bazilian, M. D., Wagner, N., & Gorini, R. (2019). The role of renewable energy in the global energy transformation. *Energy Strategy Reviews*, *24*, 38–50.

Grand View Research Incorporated (2020). Energy as a service market size worth $172.9 billion by 2027 | CAGR 14.6%: Grand View Research, Inc. *Cision PR Newswire*. Available from: https://www.prnewswire.com/news-releases/energy-as-a-service-market-size-worth-172-9-billion-by-2027--cagr-14-6-grand-view-research-inc-301014163.html.

Howarth, J. (2019). When will fossil fuels run out. *Octopus Energy*. Available from: https://octopus.energy/blog/when-will-fossil-fuels-run-out/.

Ike, I. A., Orbell, J. D., & Duke, M. (2018). Feasibility, mechanisms, and optimisation of organic pollutant degradation by thermally activated persulphate. *Chemical Engineering Research and Design*, *136*, 304–314.

Institute of Energy for South-East Europe (2020). Renewable energy market to garner 2,152.9 billion by 2025 at 4.9% CAGR, says allied market research. Available from: https://www.iene.eu/renewable-energy-market-to-garner-21529-billion-by-2025-at-49-cagr-says-allied-market-research-p4515.html.

Lehr, U., Nitsch, J., Kratzat, M., Lutz, C., & Edler, D. (2008). Renewable energy and employment in Germany. *Energy Policy*, *36*(1), 108–117.

Leonard, M. D., Michaelides, E. E., & Michaelides, D. N. (2020). Energy storage needs for the substitution of fossil fuel power plants with renewables. *Renewable Energy*, *145*, 951–962.

Lin, B., & Chen, Y. (2019). Will economic infrastructure development affect the energy intensity of China's manufacturing industry? *Energy Policy*, *132*, 122–131.

Lin, K. L., Chu, T. C., Cheng, C. J., Lee, C. H., Chang, T. C., & Wang, K. S. (2012). Recycling solar panel waste glass sintered as glass-ceramics. *Environmental Progress & Sustainable Energy*, *31*(4), 612–618.

Miao, C., Fang, D., Sun, L., Luo, Q., & Yu, Q. (2018). Driving effect of technology innovation on energy utilization efficiency in strategic emerging industries. *Journal of Cleaner Production*, *170*, 1177–1184.

Ritchie, H. & Roser, M. (2018). Energy. *Our World in Data*. Available from: https://ourworldindata.org/energy.

Shin, J., Park, J., & Park, N. (2017). A method to recycle silicon wafer from end-of-life photovoltaic module and solar panels by using recycled silicon wafers. *Solar Energy Materials and Solar Cells*, *162*, 1–6.

Steinfeld, A., & Meier, A. (2004). Solar fuels and materials. In *Encyclopedia of energy* (pp. 623–637). (Vol. 5). Elsevier Academic Press.

The National Academics o Sciences, Engineering and Medicine (2020). Energy efficient: industrial efficiency, *Need to Know*. Available from: http://needtoknow.nas.edu/energy/energy-efficiency/industrial-efficiency/.

Themelis, N. J. (2003). An overview of the global waste-to-energy industry. *Waste Management World*, 40–48.

Xu, S., Zhang, W., Li, Q., Zhao, B., Wang, S., & Long, R. (2017). Decomposition analysis of the factors that influence energy related air pollutant emission changes in China using the SDA method. *Sustainability*, *9*(10), 1742.

Young, D. C., Mill, G. A., & Wall, R. (2007). Feasibility of renewable energy storage using hydrogen in remote communities in Bhutan. *International Journal of Hydrogen Energy*, *32*(8), 997–1009.

Zhang, Q., Zhao, X., Lu, H., Ni, T., & Li, Y. (2017). Waste energy recovery and energy efficiency improvement in China's iron and steel industry. *Applied Energy*, *191*, 502–520.

Further reading

D'Agostino, D., Cuniberti, B., & Bertoldi, P. (2017). Energy consumption and efficiency technology measures in European non-residential buildings. *Energy and Buildings*, *153*, 72–86.

CHAPTER 9

Source reduction and waste minimization in textile industries

9.1 Introduction

Textile industries are one of the oldest model of industries which refer to those industries that are involved with the design, manufacturing, and distribution of cloths, dresses, and related goods. Though textile industry is one of the major chemical industries, it is necessary to understand the waste generation and waste minimization-related procedures of the textile industries in-depth, alongside the waste generation and minimization-related procedures of chemical industries. It is because textile industry is one of the most prominent industries of many developing countries that are badly struggling with adopting the source-level waste minimization for such industries. Unfortunately, textile industries are using several different harmful chemicals including heavy metals and large organic pollutants that results into highly toxic waste generation which should be reduced and managed carefully. Besides, the ever-rising growth of such industries is another factor that requires special attention in this area of source-level waste minimization in the textile industries.

In 2019, the global textile market was valued 961.5 billion USD which is estimated to experience a compound annual growth (CAGR) of at least 4.3% in between 2020 and 2027 (Ink World Magazine, 2020). At that time in 2027, as per the estimation of Grand View Research (GVR), the textile market will be worth of a total of 1,350.2 billion USD comprising the influence of rising disposable income, growing population, and non-stoppable urbanization (Grand View Research, 2020). However, the percentage of CAGR varies from country to country. For instant, among the major competitors controlling the textile industries, that is, China, United States, European Union, and the other smaller countries, the CAGR percentage varies significantly. Other competitors are also working to fast forward the growth of their corresponding textile industries. Even the countries are also working continuously to improve their corresponding situation for textile industries. For instance, India, having the status of the third largest textile manufacturing industry-consisting country, holds an approximate value of

Source Reduction and Waste Minimization
http://dx.doi.org/10.1016/B978-0-12-824320-6.00010-1

over 30 billion USD which comprises of more than 6% investment in the textile industries (Mordor Intelligence, 2020). Likewise, other developing countries also comprises significant share of the total investment in the textile industries.

However, as mentioned earlier, textile industries often deal with a variety of different fibers, dyes, colors, and other chemicals; poorly-managed disposal of such chemicals lead to different environmental and health hazards. Despite different initiatives of managing such waste, textile industry produced waste is yet one of the most hazardous groups of industrial waste. Adoption of suitable strategies in minimizing such waste involves a variety of challenges that include huge number of stakeholders involved with this industry, decentralized industries, lack of policies, and so on. Besides, many of the waste minimization technologies applicable for textile industries are not available in many parts of the world, especially in the developing countries. Moreover, many of the industry owners often have lack of interest in adopting such technologies. Thus, altogether, this sector incorporates several challenges in terms of adopting waste minimization technologies.

This chapter incorporates a discussion on some of the major type of waste produced from textile industries and at the same time, it discusses ways on minimizing these wastes at a source level.

9.2 Commonly produced waste in the textile industries: source and biogeochemical hazards

Here, textile waste denotes such type of waste that do not come into use right after the end process of the product generation. The whole process of textile industry involves a variety of individual processes which have specific type of waste produced as a by-product of the process. Such processes include spinning, weaving, knitting, dyeing, clothing, and so on. Yearly at least 10 million tons of textile waste are trashed (Sayed, 2020). It shows what a large contributor textile industry is in terms of its amount of waste generated from such industries. However, the extent of textile industry is not limited only to the waste produced directly from the textile industry, instead the clothing trashed right after the end-use are also part of textile industries. Thus altogether, textile industries have been found to be very influential contributor to waste sources. Considering the large contribution of such industries in waste generation, a variety of waste produced in the textile industries has been categorized under few major classes in Table 9.1 on the basis of waste produced at different stages of the textile industries.

Table 9.1 Commonly produced waste in the textile industries (Shahid et al., 2017; Halimoon and Yin, 2010; Agency for Toxic Substances and Disease Registry, 2013).

Waste class	Types and features	Biogeochemical hazards
Spinning waste	Spinning processes of the textile industries usually produce such waste when cotton fibers, dusts, seeds, and such waste are produced. Spinning waste may also be classified under lap waste, silver waste, roving waste, hard waste, and vacuum waste depending on the type of the waste produced at this stage.	• Light-weight fibres, dust, and other spinning waste often works as air pollutants; inhalation of these materials may rise to respiratory dysfunction. • May also hamper the lives of the aquatic species.
Weaving waste	Weaving-related processes of the textile industries usually produce such weaving waste that includes residual yarns, fibers, and other such materials. Such type of weaving waste can also be classified under sizing waste, residual waste, and some other classes depending on the characteristics of the waste.	• Since many of the spinning and weaving waste overlap the biogeochemical hazards associated with such waste also overlap. Weaving waste may also work as air and water pollutant like several spinning waste.
Knitting waste	Different types of knitting processes cause the generation of different knitting waste. Mistakes in the process leads to generation of comparatively more amount of knitting waste for which more efficient and accurate machineries are required. Besides, products wastes during trial knitting also fall under the category of waste since those trails are usually unusable.	• Knitting waste often blocks water transportation passages, hampering the easy movement of several aquatic species and thus such waste may degrade the living condition of many species.

(Continued)

Table 9.1 Commonly produced waste in the textile industries (Shahid et al., 2017; Halimoon and Yin, 2010; Agency for Toxic Substances and Disease Registry, 2013). (*Cont.*)

Waste class	Types and features	Biogeochemical hazards
Dyeing waste	Dyeing waste is the most important category of textile industry produced waste due to its extremely high toxicity and significantly high amount of generation. Textile industries use different types of dyes that usually wash away with water and thus enter into the environment. Besides, several dyeing faults also lead to the generation of waste. These wasted dyes often contain several heavy metals such as lead, chromium, cadmium, and copper. These highly toxic heavy metals enter into the environment with the textile wastewater released from the textile industries.	• Dying waste mixed with textile wastewater is one of the most prominent water pollutant that significantly degrades the lives and living condition of both aquatic and terrestrial species. • Inhalation of chromium from textile dyes may lead to the irritation of respiratory tracts causing pulmonary sensitization. Even frequent inhalation of chromium may lead to lung infection, ulcer, skin disease, nasal, or sinus cancer. • Exposure to other heavy metals present in the textile dyes also leads to other health issues.
Clothing waste	Since clothing sector of the textile industry is divided into several different segments (i.e., cutting, sewing, sorting, bundling, embroidery, finishing, etc.), the type of waste generated from such different sections also varies. However, in an ideal case, no waste was supposed to be generated from this section, but due to faults and mistakes, good share of cloths, fibers, and other materials are produced as waste.	• Random disposal of clothing waste leads to troubles in food and nutrient intake respectively for animals and plants. • With times, randomly disposed clothing waste releases dyes and other hazardous chemicals in the environment putting the lives of the species in threat.
Consumer waste	The final and one of the largest types of textile industry produced waste is consumer waste which is an end-product of the earlier-mentioned processes. Consumer waste is produced in two different ways; first right after the end use, people throw away the old clothing or due to excessive purchase, people throw away the less used cloths.	• Consumer waste in case of textile industry is one type of clothing waste mostly. Therefore, the biogeochemical hazards that the consumer waste may cause are also quite similar to the clothing waste. However, the effect of consumer waste might be even worse since such wastes are mostly disposed randomly without adopting any centralized mechanism.

Apart from, many more different classes of waste produced from the textile industries are still noticed.

9.3 Source-level waste minimization initiatives by the textile industries

Like the other industries, the source-level waste minimization initiatives in the textile industries should also follow different product substitution, technological advancement, 3R, and policymaking initiatives. Textile industry is often regarded as one of the most polluting industries due to the diverse and significant share of the waste released by such industries (da Silva et al., 2021). However, depending on the context of the textile industries, these initiatives should be changed a bit. Brief description of the suitable source-level waste minimization initiatives that can be followed in the textile industries are as followed.

9.3.1 Product substitution

Product substitution is one of the most important and necessary step to assure source level waste minimization for any of the industries. The type of products to be substituted varies from industry to industry. For example, chemical substitution should be prioritized in the case of textile chemicals, mostly metal containing dyes and mono-reactive chemicals should be replaced by more sustainable and less harmful chemicals (Shahid et al., 2017). Here, natural dyes can be considered. The synthetically produced dye utilization can be substituted by natural dye since these dyes significantly reduced the toxicity level of the waste water (Sirait, 2018). As the synthetically produced dyes often contain harsh heavy metals, substitution of these dyes will drastically reduce heavy metal containing wastewater from the textile industries.

Along with such product substitution initiatives, processes may also be substituted. For instance, as per Polprasert and Liyanage (1996), substitution of fabric based printing of the textile industry-produced cloths by thermal printing might be useful in order to reduce the chemical contents released with the textile wastewater. Even the product substitution initiatives to be adopted in the textile industries with an aim to assure source-level reduction of waste, must also focus on choosing less energy consuming products over the excessive energy consuming products (da Silva et al., 2021). Chemicals used in the textile industry may also be replaced by different enzyme utilization. It is because enzymes are highly specific and very effective in certain conditions—leading to enhanced process efficiency and technological utilization (Mojsov, 2011). Though many of the enzymes are being used by textile industry now, more and more applications of the enzyme-based procedures are needed at current time. Mostly, enzyme-based solutions can be adopted for textile product processes and biodegradation of the dye contaminants. Even cleaning, finishing, and other product processing related processes have also been found to be useful applications of enzyme based procedures (Chatha et al., 2017). Thus, product substitution might be assured replacing different chemical-based procedures by enzymatic applications.

9.3.2 Technological efficiency enhancement

Technological advancement is one of the core requirements in order to adopt effective source-level waste minimization. Center (2010) argues that cleaner production of textile industries often require incorporation of advanced technologies that should undergo sophisticated eco-efficiency assessment before adoption. In this regard, optimization of the process layout is necessary that can be as simple as energy efficient process adoption to as complex as re-circulation of textile wastewater to reuse or high pressure nozzles installation. Apart from substituting existing less-effective initiatives, more effective and efficient technological modifications may also be incorporated so that minimal amount of textile dye is wasted or minimal amount of resources are consumed to produce maximum about of goods. This indicates the urgency to optimize the less-efficient technologies for textile industrial waste minimization.

A good share of the generated textile waste is originated due to the energy inefficient procedures that cause a huge share of the chemicals to be wasted as a result of the energy inefficient processes. Such energy inefficient processes are adopted mostly due to the business-minded tendency of adopting the less expensive technologies. Though in depth cost-benefit analysis of such energy inefficient technologies determine the monetary loss driven by the energy inefficient textile technologies, knowledge gap of many textile industrialists often cause adoption of such inefficient technological approach. Among the various options of incorporation of energy efficient technologies in the textile industries, contextualization of the energy efficient technologies with the industrial context is necessary to assure profit in a low pay-back time (Hasanbeigi and Price, 2012). Such efficient technologies will play a huge role in reducing the amount of resources lost as waste in the textile industry and thus, resource recovery and minimized cost for resource purchase will be assured.

Apart from the context of energy efficiency, technological advancement in the textile industries should also focus on more and more biotechnological applications in the textile industries. Such biotechnological applications should mostly focus on the development of new type of polymers and fibers to be used in the textile industries, incorporation of enzymatic applications, effluent treatment using biotechnological approaches, and so on (Chen et al., 2007). Different biotechnological processes are often incorporated in the context of the textile industries often. But its application area should be widened. Especially, the wastewater treatment initiatives of the textile industries should have a more focus on the incorporation of biotechnology-based applications. For instance, microbial biotechnology might be used to decolorize the textile wastewater; fungus-based applications can be used for treating textile effluents; enzymatic biotechnology can be introduced in the textile industry for wet processing (Imran et al., 2015; Nielsen et al., 2009; Asgher et al., 2009). Thus, modern technology-based applications might be incorporated in different textile industries.

9.3.3 Reduce, reuse, and recycling

Textile industries can be regarded as a suitable example of 3R approach inclusion. Since this industrial setting uses a variety of products—amount of which products should be reduced at first. Initially water consumption in the textile industries should be reduced, since it has been noted that many of the textile industries are using excessive amount of water and in this regard, Shahid et al. (2017) argue that at least 10%–30% of water consumption should be reduced in the textile industries. As soon as the waste consumption will be reduced, the amount of industrial wastewater generation will also be reduced—making it easier to manage and treat the produced effluents. As mentioned in the previous section of the chapter, consumer products are one of the major type of waste produced as textile waste, reduction of consumer waste in terms of textile industrial products might be assured if everyone comes forward to reduce the amount of cloths that they consume and making the best use of the purchased cloths will also reduce the consumer produced textile waste significantly. If everyone is aware enough not to purchase more products that they need, the amount of consumer waste—as a part of textile waste—will be reduced. This is also a source level waste minimization approach since it assures minimization of consumer waste before its generation.

Second, the next methodology of the 3R approach, reuse, should also be introduced in the textile industries. In this regard, reuse of industrial wastewater should be prioritized. In order to reuse the industrial wastewater, several treatment technologies should be introduced that will not only treat the wastewater, but also recover enough resources to be used for further processing. Such processing might be assured through several different ways. Membrane-based separation is one of the commonly used wastewater treatment technology (Ciardelli et al., 2001). Such membrane-based treatment procedures are often associated with different biochemical processes in order to maximize the cleanliness of the wastewater (Bes-Piá et al., 2002). In addition to this, modern wastewater treatment technologies can also be adopted to treat the textile industry produced wastewater. For instance, Membrane Bioreactor Technology can be regarded as a promising technology in this regard, since it offers removal of contaminants based on its molecular weight (Hoinkis et al., 2012). However, conventional procedures of wastewater treatment are not often source-level waste minimization strategies for which incorporation of these treatment technologies should be incorporated as an integral part of industrial processing at the end of production. This is assured source-level waste minimization, since in this way the waste is being minimized before its exposure to the environment.

Recycling of the textile products is not as common as other industries, since in most of the cases, the waste chemicals from textile industries cannot be reused or recycled (Ütebay et al., 2020). However, many of the textile industries are still trying to recycle goods from textile industry produced waste. Therefore, the global textile waste recycling market is yet an emerging market that experience a CAGR of 19% in between 2014 and 2019 which is expected to reach at least 0.8 billion USD by 2026 with a higher CAGR of 5.2% (Piya, 2020). Even, the demand of the recycled yarns is rising gradually with several policymaking initiatives that are benefitting both the producers and consumers of the textile industries. Here, wastewater treatment technological adoption can be regarded as suitable recycling initiatives for the textile industries, since though these processes both water and valuable chemicals are recovered. Aside from this, textile recycling should also be introduced for both the pre-consumer and the post-consumer state (Hawley, 2006). In case of the pre-consumer state, the fiber-based waste produced by the textile industry can be utilized in the furniture industry for mattresses or for making other furniture (Leal Filho et al., 2019). Besides, it has been noted that agricultural waste products often work as good adsorbents in the sequestration of dyes from textile industries (Rangabhashiyam et al., 2013). Thus, the industries themselves may adopt suitable technologies in order to reduce their source-level waste at the pre-consumer state. And, the consumers can also recycle their textile products for making other useful goods or for art and craft purpose. Thus, efficient recycling of the textile processes can be assured.

9.3.4 Laws, regulations, and policies interventions

Like the other industries mentioned in the other chapters of the book, strict laws, regulations, and policies should be incorporated in order to assure source-level minimization of the textile industry produced waste. In this regard, incentives, subsidization, tax reduction, carbon tax, waste generation capacity trading mechanism, etc. can be regarded as useful ways of designing laws, regulations, and policies for source-level textile waste minimization. Besides, the government and non-government agencies should also support the textile industries until the initial payback period of waste minimization technology installation. Regulations should be imposed on the waste generation capacity of the industries. Thus, a cooperative effort of all the stakeholders is needed to assure source-level minimization of textile industry produced waste.

References

Agency for Toxic Substances & Disease Registry (2013). Chromium toxicity: what are the physiologic effects of chromium exposure. *ATSDR*. Available from: https://www.atsdr.cdc.gov/csem/csem.asp?csem=10&po=10#:~:text=Key%20Points,with%20Cr(VI)%20compounds.

Asgher, M., Azim, N., & Bhatti, H. N. (2009). Decolorization of practical textile industry effluents by white rot fungus Coriolus versicolor IBL-04. *Biochemical Engineering Journal*, *47*(1–3), 61–65.

Bes-Piá, A., Mendoza-Roca, J. A., Alcaina-Miranda, M. I., Iborra-Clar, A., & Iborra-Clar, M. I. (2002). Reuse of wastewater of the textile industry after its treatment with a combination of physico-chemical treatment and membrane technologies. *Desalination*, *149*(1–3), 169–174.

Center, L.C.P. (2010). *Cleaner production guide for textile industries*. Beirut, Lebanon: Lebanese Cleaner Production Center.

Chatha, S. A. S., Asgher, M., & Iqbal, H. M. (2017). Enzyme-based solutions for textile processing and dye contaminant biodegradation—a review. *Environmental Science and Pollution Research*, *24*(16), 14005–14018.

Chen, J., Wang, Q., Hua, Z., & Du, G. (2007). Research and application of biotechnology in textile industries in China. *Enzyme and Microbial Technology*, *40*(7), 1651–1655.

Ciardelli, G., Corsi, L., & Marcucci, M. (2001). Membrane separation for wastewater reuse in the textile industry. *Resources, Conservation and Recycling*, *31*(2), 189–197.

da Silva, P. C., de Oliveira Neto, G. C., Correia, J. M. F., & Tucci, H. N. P. (2021). Evaluation of economic, environmental and operational performance of the adoption of cleaner production: Survey in large textile industries. *Journal of Cleaner Production*, *278*, 123855.

Grand View Research (2020). Textile market size worth 1,350.2 billion USD by 2027 | CAGR: 4.3%. *Grand View Research*. Available from: https://www.grandviewresearch.com/press-release/global-textile-market.

Halimoon, N., & Yin, R. G. S. (2010). Removal of heavy metals from textile wastewater using zeolite. *Environment Asia*, *3*(2010), 124–130.

Hasanbeigi, A., & Price, L. (2012). A review of energy use and energy efficiency technologies for the textile industry. *Renewable and Sustainable Energy Reviews*, *16*(6), 3648–3665. https://doi.org/10.1016/j.rser.2012.03.029.

Hawley, J. M. (2006). Textile recycling: A systems perspective. In *Recycling in textiles*. UK: Woodhead Publishing Limited.

Hoinkis, J., Deowan, S. A., Panten, V., Figoli, A., Huang, R. R., & Drioli, E. (2012). Membrane bioreactor (MBR) technology–a promising approach for industrial water reuse. *Procedia Engineering*, *33*, 234–241.

Imran, M., Crowley, D. E., Khalid, A., Hussain, S., Mumtaz, M. W., & Arshad, M. (2015). Microbial biotechnology for decolorization of textile wastewaters. *Reviews in Environmental Science and Bio/Technology*, *14*(1), 73–92.

Ink World Magazine (2020). Global textile market to register CAGR of 4.3% from 2020 to 2027: Grand View Research. *Ink World Magazine*. Available from: https://www.inkworldmagazine.com/contents/view_breaking-news/2020-02-19/global-textile-market-to-register-cagr-of-43-from-2020-to-2027-grand-view-research/#:~:text=02.19.20-,The%20global%20textile%20market%20size%20was%20valued%20at%20%24961.5%20billion,according%20to%20Grand%20View%20Research.

Leal Filho, W., Ellams, D., Han, S., Tyler, D., Boiten, V. J., Paço, A., et al. (2019). A review of the socio-economic advantages of textile recycling. *Journal of Cleaner Production*, *218*, 10–20.

Mojsov, K. (2011). Application of enzymes in the textile industry: a review. In: *Proceedings of the eleventh international congress "engineering, ecology and materials in the processing industry,"* (pp. 230–239).

Mordor Intelligence (2020). Global textile industry—growth, trends, covid-19 impact and forecasts (2021–2026). *Mordor Intelligence*. Available from: https://www.mordorintelligence.com/industry-reports/global-textile-industry---growth-trends-and-forecast-2019---2024.

Nielsen, P. H., Kuilderd, H., Zhou, W., & Lu, X. (2009). Enzyme biotechnology for sustainable textiles. In *Sustainable textiles* (pp. 113–138). Woodhead Publishing.

Piya, F.A. (2020). Textiles waste and recycling options. *Textile Focus*. Available from: http://textilefocus.com/10919-2/.

Polprasert, C., & Liyanage, L. R. J. (1996). Hazardous waste generation and processing. *Resources, Conservation and Recycling*, *16*(1–4), 213–226.

Rangabhashiyam, S., Anu, N., & Selvaraju, N. (2013). Sequestration of dye from textile industry wastewater using agricultural waste products as adsorbents. *Journal of Environmental Chemical Engineering*, *1*(4), 629–641.

Sayed, M.A. (2020) What is textile waste | definition, meaning and types. *Textile Apex*.

Shahid, A.M., Sela, S.K., Hossain, N.A.K.M. (2017). Waste minimization in textile industry. *Textile Today*. Available from: https://www.textiletoday.com.bd/waste-minimization-textile-industry/.

Sirait, M. (2018). Cleaner production options for reducing industrial waste: the case of batik industry in Malang, East Java-Indonesia. In *IOP conference series: earth and environmental science* (Vol. 106, No. 1, p. 012069). IOP Publishing.

Ütebay, B., Çelik, P., & Çay, A. (2020). Textile wastes: status and perspectives. In *Waste in textile and leather sectors*. IntechOpen.

CHAPTER 10

Source reduction and waste minimization in the mining industries

10.1 Introduction

Mining industry is one of the most prominent earning source of many different countries, since the growth of the mining industries often regulate the resource acquisition potential and economic growth of the countries. Based on different characteristics of the mining industries, this industry might be categorized under geophysical industries or chemical industries. However, though mining industry is a very less-diverse and small industrial setting, understanding the waste management initiatives or exploring initiatives in order to minimize mining-generated waste at source level, are necessary to tackle any accidents or to protect the surround environment and environmental components. Mining industry indicates the cluster of process that are involved with extraction, management, and processing of naturally occurring solid minerals from the earth surface. As a product of mining, various economically valuable products might be obtained, that is, coal, diamond, metallic ores, oil, and so on. Mining industry also involve processes associated with mineral processing, surface mining, market of mining equipment, etc. Sustainability of the mining industries is largely correlated with its corresponding industrial impacts on the environment, since there are enough evidences of the detrimental impacts of mining industry on environment (Hilson and Murck, 2000).

Despite the negative impacts that mining industry may bring about, there is no scope to bypass the positive outcome that this industry assures. Despite being a very small-scale industry with quite less interaction with other industries, it is noted that in 2019, the global mining equipment market size has been found to be 1,21,694.3 million USD that is predicted to be at least 1,65,827.8 million USD by 2027 with a potential compound annual growth rate (CAGR) of 5.7% (ReportLinker, 2020). Looking at the potential market size of mining industry,

Source Reduction and Waste Minimization
http://dx.doi.org/10.1016/B978-0-12-824320-6.00011-3

the impact of mining industry in the world economy is easily understandable. Many of the mined goods are used for energy production and used as raw materials for chemical industries which also let the mining industry control the growth of other industries such as energy production industry, chemical industry, electrical industry, electronics industry, and so on.

However, mining industry is a very old industry due to the age-old anthropogenic dependency of the human being on the natural resources. Since the Stone Age, mining industry has developed and gradually adopted today's shape of the mining industry. The potentials and capacity of mining also help a country to control the growth of their economy. For instance, countries of the Middle East, have been successful to extract oil and thus, they have achieved tremendous economic growth over the years. Only existence of geological resources is not enough to assure economic growth, if innovative and modern inclusions are not assured alongside the contemporary processes of mining. For example, with the aid of the government of China, the country could promote their mining industry; especially coal extraction, utilizing many innovative ideas (Fan et al., 2017). However, there are many challenges associated with this mining industry that includes, lack of sustainable approaches, need for expensive and advanced technologies, geo-physical or climatic variables, and so on. Besides, the huge amount of waste produced from this industry is also another big challenge to be tackled in this sector.

Therefore, considering the huge amount of waste produced from such a leading industry, mining industry, this chapter aims to explore the major types of waste produced from mining industries and at the same time, this chapter also suggests suitable approaches to be adopted in order to assure source-level minimization of mining industry-produced waste.

10.2 Commonly produced waste in the mining industries: source and biogeochemical hazards

The mining industries produce a variety of wastes which have been categorized below along with their corresponding examples and biogeochemical hazards. This chapter classifies waste produced in the mining industry as per its state in the normal temperature that makes three major classes of waste, solid waste, liquid waste, and gaseous waste. Additionally, there are some special characterized classes of waste that might also be produced in some mining industries (Table 10.1).

Table 10.1 Commonly produced waste in the mining industries (Karu et al., 2013).

Waste class	Types and features	Biogeochemical hazards
Solid waste	Solid waste produced in the mining industries include, gravels, clay, sands, rocks, metallic ores, metal particles, coals, sands, and so on. These solid wastes are mainly produced as a byproduct of the excavation process. Besides, based on the product that is being mined, the type of solid waste also varies; especially the ore of the mined products are found as a solid waste. The type of solid mining waste also depends of the technologies used that vary from country to country.	• Metallic ores or coal components enter into the food cycle putting the lives of the animals in a great threat. • Promotes water and air pollution.
Liquid waste	Liquid wastes produced in the mining industries are mostly either the chemicals used in the process of mining or the products extracted in the process of mining. For instance, in oil mining plants, oil is a major type of waste produced in the process of mining. Metallic ores are often mixed with the nearby water source causing water pollution. Many of the mining industries use cyanide as an extracting chemical which is highly toxic to human. It has been noted that cyanide used in the mining plants are often released with the waste from such plants. Such water is commonly known as mine water that is produced as a by-product of the mining industries. Besides, the concept of acid mine drainage is necessary to be understood in this regard. Acid mine drainage is the process when water flows in such a pathway which is rich in sulfur content and thus during its flow it produces acidic solutions; which is quite common in the surrounding area of coal mining.	• Cause water pollution • Initiate the deaths of the aquatic animals and plants when the type of the waste is highly hazardous • Pollutes drinking water source. • Oil spilt in water cause less light penetration and thus kills aquatic plants. • Hazardous chemicals may be ingested by human from water or other food sources that may cause several health issues or even death. • Acid Mine Drainage (AMD) leads to the death of the affected animals and plants.

(*Continued*)

Table 10.1 Commonly produced waste in the mining industries (Karu et al., 2013). (*Cont.*)

Waste class	Types and features	Biogeochemical hazards
Gaseous Waste	Mining industries also produce good amount of air pollutants, usually gaseous waste, that instantly get mixed with the air and pollutes the surrounding environment. This includes particulate materials mostly. PM2.5 and PM10 materials are often produced in the mining industries which are mainly residues of the extracted materials and the process chemicals. Aside from these, mining industry produces other air pollutants which are not gaseous in normal temperature but these are highly responsible for air pollution of the surrounding environment. Example of such type of waste includes dust, ashes, sand particles, and so on.	• Gaseous wastes are released as harmful chemicals in the air which often cause dangerous health issues. • Huge amount of air pollutants released from such mining plants cause the death of nearby plants and death or migration of other animals and insects. • Gaseous wastes produced from the mines are often spread at a longer range of area causing harm in the human habitats. • Radioactive wastes which are produced from the mining industries are often responsible for causing very bad health hazards that may include cognitive impairment, cancers, and other health issues. • Release of such various wastes from mining industry also makes the nearby area inhabitable.
Other important classes of waste	Apart from the earlier-mentioned waste classes, mining industries often incorporates generation of other classes of waste that includes radioactive waste, carcinogenic waste, and other classes of highly toxic waste. Generation of such waste is dependent on the type of the mining plants. For instance, the process of uranium mining, often release radioactive uranium as waste. Besides, many carcinogenic chemicals are often used in the process of mining that are often released to the environment due to the poor management of the waste. Apart from these, there are potentials of the generation of other types of waste depending on the type of the mining industry.	

10.3 Source-level waste minimization initiatives by the mining industries

As mentioned in the previous segment of the chapter, mining industries produce a variety of wastes which are very poorly managed in most of the cases. Besides, due to the smaller extents of most of the mining industries, there are very less scope to introduce waste management plants connecting with the mining industries. Therefore, adopting source-level waste minimization will be a very suitable approach in case of the mining industry, since small scale mining industries will be capable of incorporating such initiative more effectively than the postgeneration waste management initiatives. Adoption of mining waste reduction initiatives will protect the environment drastically, since mining activities is one of the major environmental problems caused by anthropogenic activities in certain parts of the world. Though the overall process of mining waste minimization involves several economic, legislation, and technological challenges, there are chances yet to implement pollution prevention-aimed strategic approaches (Hilson, 2000). However, like other chapters, some of the major source-level waste minimization initiatives to be adopted by the mining industries have been discussed as followed.

10.3.1 Product substitution

Mining industry is highly correlated with the mineral and construction-material producing industries, since in many cases mining activities assist in these industries. Therefore, product substitution, as a source level waste minimization initiative, for mining industrial setting have also incorporated evidence of different mineral and construction-material producing industries. Like other industries discussed in this book, product substitution for mining industries also aims a shift from less sustainable products to more sustainable products that will also reduce the amount of waste generated. For instance, natural clay can be substituted by calamine processing waste while manufacturing value added fired bricks (Taha et al., 2016). This is a suitable waste minimization initiative since it offers both utilization of waste and minimization of natural clay-induced waste generation. At this process, useful and valuable goods, natural clay, is also being saved. However, the process of product substitution is not always much smooth, since in many cases substitution to sustainable products are highly expensive, that discourages many industrialists to shift toward these products (Humphreys, 2001). Then the necessity to design affordable but sustainable products through innovations and research is required.

10.3.2 Technological efficiency enhancement

The existing technology of the mining industries should be upgraded with an aim to adopt more efficient and sustainable technology. When the technologies used for a certain process will be efficient enough, there will be comparatively less amount of resources misused and disposed with the waste. This denotes that efficient technologies will help to save enough resource, thus it will reduce the amount of waste generated from the process. In most of the cases, the amount of waste generated from the mining industry affects the surrounding environment in many different ways. Therefore, advanced technologies should be installed in order to minimize its impacts on the surrounding environment. The purpose of such technological shift is to move forward sustainability. Though this technological shift often incorporates challenges, the industry owners should be encouraged to adopt these technologies (Hilson and Murck, 2000). In this regard, the adopted technologies might be for sustainable extraction, processing, or management.

However, suitable technological upgradation in this regard might include carbon capture technologies, advanced filtration, or associative treatment plants. As a carbon capture rock, suitable mineral-based rocks can be installed in the coal mining or carbon di- or mono-oxide producing plants. It has been found that suitably designed mineral rocks can be used to capture and store produced carbon containing cases from the mining plants (Hitch et al., 2010). Apart from that advanced filtration mechanism should also be incorporated with the mining plants. Several different membranous or filtration technologies can be adopted in this regard. For example, hydrate gel membrane technology can be used for the filtration of the mine tailings which process actually thickens the waste suspension making it not permeable through certain membrane. Thus, only the expected products pass through the semi-permeable membrane causing a sharp reduction to the unwanted waste disposal to the environment (Liu et al., 2019). Later, based on the types and classes of the isolated solid materials, the isolated waste can be used in multiple different sectors as a raw material or can be refined for resource recovery purpose. There are many more membrane-based filtration systems that can be incorporated in the mining industries to isolate its unwanted waste.

Apart from that, treatment plants can also be installed along with the mining industries that will assist in minimizing the generated waste at source. Though installation of treatment plants is not a very good source-level waste reduction mechanism, installation of the treatment facilities at the waste source is a better option than many other postgeneration waste management initiatives. These treatment plants should be designed with the inclusion of several different biochemical approaches so that all the different type of chemical waste can be isolated from the waste cluster (Huisman et al., 2006). Apart from that, a special attention should be given to the reduction of Acid Mine Drainage (AMD), since AMD is one of the most common form of mining-based pollution. Though biochemical process is the most suitable initiative for AMD reduction, installation of lagoons or wetland may also reduce the extent of the exposure reduction for AMD (Akcil and Koldas, 2006). Thus, suitable treatment plants should be installed in reducing the amount of waste generated from the mining industries.

10.3.3 Inclusion of 3R approach

There is not much scope for the applications of 3R approach in the mining industry since this industry is mostly focused on resource extraction. However, the corresponding industrialists should focus on reducing the misuse of products. It is not necessary only to protect our environment from pollution, but also to reduce the generation cost of the industrialists. Aside from this, reuse of the products should be assured. A good number of ores are wasted in the mineral, mining, and construction industries which can be easily extracted from the waste cluster and thus this should be reused (Yellishetty et al., 2008). If possible, recycling initiatives should also be incorporated in such regard. Recycling approaches can be incorporated in multiple ways. For instance, coal mining industries of China are recycling the mine waste to extract valuable goods out of it (Haibin and Zhenling, 2010). Even, it has been noted that limestone industries can use their produced waste for concrete production (Rana et al., 2017). Thus, based on the applicable contexts, suitable reduction, reuse, and recycling strategies can be adopted in different sectors of the mining industries.

10.3.4 Laws, regulations, and policy interventions

Like other industries, source-level waste minimization in the mining industries should also incorporate several different laws, regulations, and policies. The government should initiate subsidization for the adoption of source-level waste minimization mechanism. Incentives, tax reduction, and other facilities should also be offered to respective industries to encourage their participation in adopting suitable mining waste minimization initiatives. Environmental monitoring mechanism is needed to be imposed strictly in the mining industries (Gorova et al., 2013). Even well-managed source-level waste minimization in the mining industries can play an important role in the circular economy by adding value of the generated mining waste in producing useful goods and by reducing the corresponding environmental hazards (Lèbre et al., 2017). Thus, through the initiatives of all different stakeholders, a sustainable source-level waste minimization approach can be assured for the mining industry.

References

Akcil, A., & Koldas, S. (2006). Acid Mine Drainage (AMD): causes, treatment and case studies. *Journal of Cleaner Production, 14*(12–13), 1139–1145.

Fan, S., Yan, J., & Sha, J. (2017). Innovation and economic growth in the mining industry: Evidence from China's listed companies. *Resources Policy, 54*, 25–42.

Gorova, A., Pavlychenko, A., & Borysovs'Ka, O. (2013). The study of ecological state of waste disposal areas of energy and mining companies. *Mining of mineral deposits*. CRC Press. 169-171.

Haibin, L., & Zhenling, L. (2010). Recycling utilization patterns of coal mining waste in China. *Resources, Conservation and Recycling, 54*(12), 1331–1340.

Hilson, G. (2000). Pollution prevention and cleaner production in the mining industry: an analysis of current issues. *Journal of Cleaner Production, 8*(2), 119–126.

Hilson, G., & Murck, B. (2000). Sustainable development in the mining industry: clarifying the corporate perspective. *Resources Policy, 26*(4), 227–238.

Hitch, M., Ballantyne, S. M., & Hindle, S. R. (2010). Revaluing mine waste rock for carbon capture and storage. *International Journal of Mining, Reclamation and Environment, 24*(1), 64–79.

Huisman, J. L., Schouten, G., & Schultz, C. (2006). Biologically produced sulphide for purification of process streams, effluent treatment and recovery of metals in the metal and mining industry. *Hydrometallurgy, 83*(1–4), 106–113.

Humphreys, D. (2001). Sustainable development: can the mining industry afford it? *Resources Policy, 27*(1), 1–7.

Karu, V., Valgma, I., & Rahe, T. (2013). Mining waste reduction methods. In *Thirteenth international symposium " topical problems in the field of electrical and power engineering."* Doctoral Scholl of Energy and Geotechnology II, Pärnu, Estonia (pp. 14–19).

Lèbre, É., Corder, G., & Golev, A. (2017). The role of the mining industry in a circular economy: a framework for resource management at the mine site level. *Journal of Industrial Ecology, 21*(3), 662–672.

Liu, D., Edraki, M., Malekizadeh, A., Schenk, P. M., & Berry, L. (2019). Introducing the hydrate gel membrane technology for filtration of mine tailings. *Minerals Engineering, 135*, 1–8.

Rana, A., Kalla, P., & Csetenyi, L. J. (2017). Recycling of dimension limestone industry waste in concrete. *International Journal of Mining, Reclamation and Environment, 31*(4), 231–250.

ReportLinker (2020). Mining equipment market analysis by type and application: global opportunity analysis and industry forecase, 2020-2027. GlobeNewsWire. Available from: https://www.globenewswire.com/news-release/2020/09/02/2087474/0/en/The-global-mining-equipment-market-size-was-valued-at-121-694-3-million-in-2019-and-is-expected-to-reach-165-827-8-million-by-2027-growing-at-a-CAGR-of-5-7-from-2020-to-2027.html. Accessed on 1 December 2020

Taha, Y., Benzaazoua, M., Hakkou, R., & Mansori, M. (2016). Natural clay substitution by calamine processing wastes to manufacture fired bricks. *Journal of Cleaner Production, 135*, 847–858.

Yellishetty, M., Karpe, V., Reddy, E. H., Subhash, K. N., & Ranjith, P. G. (2008). Reuse of iron ore mineral wastes in civil engineering constructions: A case study. *Resources, Conservation and Recycling, 52*(11), 1283–1289.

CHAPTER 11

Challenges and recommendations for source reduction and waste minimization

11.1 Introduction

Source reduction and waste minimization encourages prevention of waste before its generation. As discussed in all the previous chapters, it is a very safer approach especially for industrial waste minimization. Since it prevents toxic waste generation, the corresponding hazards can be reduced significantly, as under a proper waste source reduction scheme, both the toxicity and the amount of hazardous waste generation are minimized. However, there are several challenges associated with the issue that makes it even more difficult to assure large scale implementation. The type of challenges faced by the industries of different geographical context is also very diverse in nature as per the industrial set up, geographic context, business categories, cultural beliefs, and other (Bai and Sutanto, 2002; Morone et al., 2019). All these different challenges make the implementation of the industrial waste source reduction initiatives even tougher. However, adoption of suitable strategies may improve the situation drastically. In doing so, the responsibilities of all the sectors are important and they should come forward to play their individual responsibilities.

This chapter summarizes few major challenges that might be commonly faced during waste source reduction and also suggests potential strategies to overcome these challenges.

11.2 Corresponding challenges and recommendations for source reduction and waste minimization

Among many of the challenges associated with industrial waste source reduction; some of the common challenges are industrial reluctance, technological/financial limitation, insufficient training/knowledge dissemination, and so on. They along with some corresponding strategies have been discussed later with corresponding strategies to overcome these.

Source Reduction and Waste Minimization
http://dx.doi.org/10.1016/B978-0-12-824320-6.00009-5

11.2.1 Industrial reluctance

Industrial reluctance can be regarded as the most responsible factor that hinders source-level waste minimization initiative adoption. Most of the industries are truly vast industries with numerous numbers of smaller to larger role players. These players often do not coordinate among themselves to adopt suitable waste minimization initiatives. Instead, they are found to focus more on profit making. However, their tendency of profit making will not lead to long term sustainability for them and eventually it will lead to their corresponding industries to collapse (Al-Hajj and Hamani, 2011). It is because, failure to adopt waste minimization approaches often leads to unavoidable accidents and hazards causing the industrial role players spend much on waste-borne hazard minimization for a long period of time. Additionally with the increasing safety concerns of the customers, the customers are also getting more concerned about the safety, cleanliness, and consequences of industrial hazards. The growing eagerness of the customer to organic and cleaner industrial products can be regarded as an indicator of how most of the customers are going to react toward less clean and more waste producing industrial products in near future (Prentice et al., 2018). Even many people, old to young, often prefer reusable substances over one-time products considering the corresponding cost they need to pay for disposable and reusable products. Market-based researches have also showed success in terms of consumer's positive behavior toward reusable goods (Reicks et al., 1997). If any industry keeps being reluctant in this context, eventually they might be boycotted by the consumers causing collapse of the industries. Considering all these, the reluctance of the industries should be changed, if they want to sustain in the market. The industries should incorporate market study and market assessment mechanism in their industrial supply chain, if they want to upgrade or improve their durability and sustainability in the market.

Reluctance of the industries is also associated with several other concerns. For example, in many of the countries, the government has not introduced enough facilities or policies to assure industrial involvement toward waste minimization. Sometimes, even if there are suitable policies adopted, the monitoring and evaluation mechanism is so poor that the industries often do not feel the need to prevent waste generation. Though such policy adoption often encourage many of the industries to adopt suitable policies, unfortunately poor monitoring of the monitoring authority often cause reluctance of the industrial role-playing bodies to maintain, operate, and regulate the installed facilities properly (Tsai and Chou, 2004). In addition, many of the smaller industries are often not considered under the adoption of such policies. As a result, these industries are not concerned to minimize their produced waste at all. Therefore, their small contribution is tremendously multiplied because of having numerous numbers of smaller role-players. To overcome such relucting tendency of these smaller industries, policies should be adopted in such a way so that all different industries; smaller or larger, contribute in source-level waste minimization from their respective contexts (Szarka, 2006). Industrial reluctance can also be caused due to many more reasons that should be carefully identified and overcome through more initiative adoption.

11.2.2 Inadequate training facilities

In addition to the industrial reluctance, insufficient contextualized training and educational facilities are also responsible for inadequate industrial initiatives for source-level industrial waste minimization. Inadequate training opportunities is a major challenge for sustainable source-level waste minimization, since if the industrial employees are not trained and educated properly in terms of their corresponding responsibilities of source-level waste minimization, they won't be able to design source-level waste minimization sustainably. Gakungu et al. (2012) argue that when the employees do not receive proper contextualized training under experience trainers, if they are forced to design waste minimization initiatives due to industrial and government regulations, the designed mechanisms of the non-trained employees are mostly collapsed within a short duration of time. Such evidences are more prevalent in the low income developing countries since their training opportunities are highly inadequate (Warner, 2016). Even the earlier mentioned source-level waste minimization challenge, industrial reluctance, is often caused a result of inadequacy of training and educational facilities for the respective sectors. This situation is also vice-versa, since industrial reluctance also initiate inadequate training and educational facilities for the industrial employees. Even with such limitations, many of the industrial workers are not concerned about their roles and responsibilities in terms of source-level industrial waste minimization for whom, private firms are often found to provide good quality waste minimization training facilities. Such private initiatives should also be promoted by the industries and also by government regulations.

Inadequate training facilities limit implementation of source-level waste minimization initiatives, since it keeps the employees ignorant of their role and necessities regarding waste minimization initiatives. Industries that could successfully install waste minimization initiatives have been found to constantly train their employees and also required regular upgrade and modifications of the installed waste minimization technologies and strategies. In this regard, the role of the government is also important. Many of the developed countries have been found to arrange training for their industrial employees along with the commercial or private training initiatives (Phillips et al., 2003). In such cases, all the training should also be contextualized as per the geophysical contexts and industrial set up. For instance, though both pharmaceutical and commercially needed chemical industries are found to use similar type of processes and often have several similarities in terms of their produced waste, the minimization strategy should not be the same (Wang et al., 2015). It is so because there are numerous scopes of cross-contamination or intrusion of highly hazardous chemicals. For instance, pharmaceutical industries often use several carcinogenic compounds that may get mixed with other types of waste for which more safety concerns should be incorporated in case of pharmaceutical industries. However, such initiatives should not be limited only to these, instead, intense analysis of all the chemicals and cross-contamination analysis pathways should also be analyzed properly. And in doing so, the employees should be trained properly as per the initiatives of their corresponding industries and government bodies so that they can modify the system as needed.

11.2.3 Technological limitations

Technology can be regarded as the major drive of concurrent world. Technological initiatives determine the efficiency of industrial processes and also the growth tremendously. However, industries should also be careful to install suitable waste minimization initiatives. Unfortunately, many of the industries are struggling to do so. The expenses of the technologies are often quite high and in many cases, the technologies are not easily found everywhere. For instance, the industries that are trying to incorporate the waste minimization initiatives, often fail to find out any suitable technologies other than incineration. Since incineration doesn't need any complex set up or advanced technological components, it is a very popular postgeneration waste minimization strategy, especially for organic waste (Allsopp et al., 2001). Though there are other suitable options to minimize such waste, these are not that common due to the technological complexities and unavailability of several technologies. Especially adequate amount of advanced waste minimization technologies are quite rare in the low income countries, since all the industrialists cannot often afford the technologies. Even if they can afford, their reluctance prohibit them from using such technologies. For instance, though a good number of technological initiatives have already been discovered for efficient metal extraction, many of the countries are still not being able to utilize those facilities, since the market didn't prefer to sell those due to lack of affordability of the potential stakeholders and also for industrial reluctance (Pant et al., 2012; Cheremisinoff, 2003). Thus, sometimes, suitable technologies for industrial applications have not yet been innovated or sometimes these technologies are not available at all different locations. Considering such issues, technological limitations can be regarded as a major challenge for industrial waste minimization that might be overcome through more innovative researches or through ensuring broader availability of waste minimization initiatives.

11.2.4 Financial limitations

Financing the waste minimization initiatives is also another major challenge that should be overcome so that the industries can install initiatives for source-level waste minimization. The smaller industrial role players cannot often afford to install source-level waste minimization technologies. Therefore, laws and policies should be implemented by governing agencies so that the financial limitations of certain industries can be overcome. Subsidization can be regarded as a suitable strategy in this regard. While determining the amount of subsidization, some of the potential factors to be considered, might be the industries' financial condition, amount and types of produced waste, availability of industrial co-current waste minimization facilities, and so on. Government subsidization has tremendously helped to grow industrial initiatives for waste minimization in countries like Taiwan (Young et al., 2010). Along with these, incentives, tax reduction, and other financial supports can also be ensured for the industries so that they become interested to adopt source-level waste minimization strategies.

Along with these, sometimes geographic barriers, cultural beliefs, etc. might also be challenging for source-level waste minimization strategy implementation at some contexts. To overcome such challenges, the context should be critically analyzed and researched in depth. Thus, some of the major challenges faced during source-level waste minimization strategies implementation can be overcome.

References

Al-Hajj, A., & Hamani, K. (2011). Material waste in the UAE construction industry: Main causes and minimization practices. *Architectural Engineering and Design Management*, 7(4), 221–235.

Allsopp, M., Costner, P., & Johnston, P. (2001). Incineration and human health. *Environmental Science and Pollution Research, 8*(2), 141–145.
Bai, R., & Sutanto, M. (2002). The practice and challenges of solid waste management in Singapore. *Waste Management, 22*(5), 557–567.
Cheremisinoff, N. P. (2003). *Handbook of Solid Waste Management and Waste Minimization Technologies*. Butterworth-Heinemann.
Gakungu, N. K., Gitau, A. N., Njoroge, B. N. K., & Kimani, M. W. (2012). Solid waste management in Kenya: a case study of public technical training institutions. *ICASTOR Journal of Engineering, 5*(3), 127–138.
Morone, P., Koutinas, A., Gathergood, N., Arshadi, M., & Matharu, A. (2019). Food waste: challenges and opportunities for enhancing the emerging bio-economy. *Journal of Cleaner Production, 221*, 10–16.
Pant, D., Joshi, D., Upreti, M. K., & Kotnala, R. K. (2012). Chemical and biological extraction of metals present in E waste: a hybrid technology. *Waste Management, 32*(5), 979–990.
Phillips, P. S., Clarkson, P., Adams, J., Read, A. D., & Coggins, P. C. (2003). County waste minimization programmes: a case study from Northamptonshire, UK. *Sustainable Development, 11*(2), 103–118.
Prentice, C., Wang, X., & Lin, X. (2018). An organic approach to customer engagement and loyalty. *Journal of Computer Information Systems, 60*, 326–335.
Reicks, M., Splett, P., & Fishman, A. (1997). *Shelf labeling of organic foods: effects on customer perceptions and sales* (No. 1710-2016-140019).
Szarka, J. (2006). From inadvertent to reluctant pioneer? Climate strategies and policy style in France. *Climate Policy, 5*(6), 627–638.
Tsai, W.T., & Chou, Y. H. (2004). Government policies for encouraging industrial waste reuse and pollution prevention in Taiwan. *Journal of Cleaner Production, 12*(7), 725–736.
Wang, J., Li, Z., & Tam, V. W. (2015). Identifying best design strategies for construction waste minimization. *Journal of Cleaner Production, 92*, 237–247.
Warner, M. (2016). *How Chinese managers learn: Management and industrial training in China*. Springer.
Young, C. Y., Ni, S. P., & Fan, K. S. (2010). Working towards a zero waste environment in Taiwan. *Waste Management & Research, 28*(3), 236–244.

Index

Note: Page numbers followed by "f" indicate figures, "t" indicate tables.

A

Acid mine drainage (AMD), 174
Alternative energy sources, 45
Anaerobic digestion, 27
Appliance recycling, 77
Automatic waste collection, 47

B

Biofiltration, 40
Biogeochemical hazards, 63
Biological treatment, 24, 26
  for metal extraction, 72

C

CAGR. *See* Compound annual growth rate (CAGR)
Carbon filtering, 40
Case studies
  Finelite Inc.'s Effort to Reduce Waste, 17
  food waste recycling—South Korea's effort to become a zero waste society, 52
  reconomy's effort to incorporate waste management initiatives in business sectors, 124
  sustainable medical waste minimization plan by "reduction in motion", 106
  waste to resource concept implementation through metal recovery mechanisms, 77
  waste to wealth—utilization of coconut shell for plantation, 51
Chemical fixation, 74
Chemical industries, 131
  gaseous chemical waste, 130
  laws, regulations, and policies interventions, 133
  liquid chemical waste, 129
  prerelease treatment, 133
  process modification, 131
  product substitution, 131
  3R approach, 132
  solid chemical waste, 129
  source-level waste minimization initiatives, 131
  technological advancement, 131
Chemical treatment, 24, 31
  for metal extraction, 72
Cleaner technologies, 23
Co-benefits of waste source reduction and waste minimization
  economic benefits, 15
  environmental co-benefits, 12
  health-related co-benefits, 13
Code of Federal Regulations (CFR), section 261, 4
Composting, 24, 25
Compound annual growth rate (CAGR), 61, 83, 137
Construction industry, 111, 112
  commonly produced waste and biogeochemical hazards, 113, 115t
  waste source reduction mechanism, 118f
  source-level waste minimization initiatives by, 114
    landfilling, 114
    laws, regulations, and policies interventions, 123
    product substitution, 119
    process modifications, 119
    3R approach, 121
    technological advancement, 119
Copper recovery, through chemical leaching, 73

E

Electrical and electronics industry, 61
  commonly produced waste in, 63
  source-level waste minimization initiatives by, 69
    process modification, 70
    recycling, 71
  waste source reduction strategies in, 70f
Electrical and electronics industry generated waste, 4
Electrical devices, 61

Electrokinetic remediation, 73
Electrolysis, 32
Electronic devices, 61
Energy producing industry, 151
- commonly produced waste, 152t
- laws, regulations, and policies interventions, 156
- product substitution, 151
- recycling and reusing, 155
- source-level waste minimization initiatives, 151
- technological efficiency enhancement, 154

Energy source modification, 24, 44
Environmental Protection Agency (EPA), 4
Enzyme-based waste minimization, 28
E-waste, 63
- laws, regulations and policies to minimize source-level, 79t
- metallic e-waste, 64
  - sources and hazards, 64t
- organic e-waste
  - reduction, reuse, and recycling, 77
  - sources and hazards, 68t

F

Filtration, 24, 39
Financial limitations, 181
Food industry, 137
- commonly produced waste, 139, 140t
- laws, regulations, and policies interventions, 146
- process modifications, 143
- 3R approach, 144
- source-level waste minimization initiatives by, 139

Food wastes, 139

G

Government subsidization, 181
Green chemistry, 48
- principles of, 49t

Green house gases (GHGs) emission, 16, 150

H

Hazardous waste materials, 4
HCW. *See* Healthcare waste (HCW)
Healthcare industry, 83
- commonly produced waste and biogeochemical hazards, 85
- major classes of commonly produced waste in, 86t
- source-level waste minimization initiatives, 85

Healthcare industry infectious waste, 4
Healthcare waste (HCW), 83
- biochemical treatment, 102
- composting, 99
- incineration, 102
- laws, regulations, and policies intervention, 103
- process modification, 93
- product substitution, 93
- reducing, reusing, and recycling, 98
- technological advancement, 93
- thermal treatment, 101
- waste segregation, 91

Hydrothermal carbonization, 37
Hydrothermal liquefaction (HTL), 38

I

Inadequate training facilities, 179
Incineration, 34
Industrial reluctance, 178
Industrial waste source reduction, challenges associated with, 177
- financial limitations, 181
- inadequate training facilities, 179
- industrial reluctance, 178
- technological limitations, 180

Information technology-based waste minimization, 46
Integrated waste management strategies, 10
- technical and non-technical components of, 10f

L

Laws, regulations and policies to minimize source-level e-waste, 79t
Liquid waste management, 8, 9f

M

Membrane-based system, 30
Metal extraction from e-waste, 71

Metallic e-waste, 64
  sources and hazards, 64t
Metallic waste, 4, 64
Mining industry, 169
  commonly produced waste in, 170, 171t
  inclusion of 3R approach, 175
  laws, regulations, and policy interventions, 175
  product substitution, 173
  source-level waste minimization initiatives, 173
  technological efficiency enhancemen, 174
Molten salt oxidation, 36

N
Non-renewable energy production industry, 149

O
Ocean cleanup system, 43
Organic e-waste reduction, reuse, and recycling, 77
Organic e-waste sources and hazards, 68t
Organic waste, 67

P
Physical treatment for metal extraction, 72
Plasma gasification system, 39
Process modification, 70, 93, 131, 143
Pyrolysis, 33

R
3R (reduce, reuse, and recycle) concept, 5
3R model, basic workflow of, 6f
Recycling, 24, 71, 77, 155, 165
Renewable energy generation, 44
Reverse osmosis, 42
RFID (radio frequency identification) tracking system, 46

S
SDG. *See* Sustainable development goals (SDGs)
Solar energy, 155
Solid waste management, 6, 7f
Sustainable development goals (SDGs), 16, 23
Sustainable waste minimization strategies, 5

T
Technological limitations, 180
Textile industries, 159
  commonly produced waste in, 160, 161t
  laws, regulations, and policies interventions, 167
  product substitution, 163
  reduce, reuse, and recycling, 165
  source-level waste minimization initiatives, 163
  technological efficiency enhancement, 164
Thermal treatment, 24, 34

U
Ultrafiltration, 41

W
Waste autoclave, 30
Waste-borne hazards, 23
Waste from electronic and electrical equipment (WEEE), 62
Waste minimization, 3
  and source reduction strategies, 49t
  and sustainable development goals (SDGs), 16
Waste minimization strategies, 23, 48. *See also specific entries*
Waste transportation, 23

Z
Zero waste strategy, 11

Kolibri 360 GmbH
Ferdinand-Jühlke-Straße 7
99095 Erfurt, Deutschland
produktsicherheit@kolibri360.de